A Woman's Way

A Woman's Way

The Forgotten History of Women Spiritual Directors

Patricia Ranft

palgrave

ISBN 978-1-349-42578-5   ISBN 978-0-312-29947-7 (eBook)
DOI 10.1057/9780312299477

A WOMAN'S WAY: THE FORGOTTEN HISTORY OF WOMEN SPIRITUAL DIRECTORS
Copyright © Patricia Ranft, 2000.
Softcover reprint of the hardcover 1st edition 2000 978-0-333-92989-6

First published 2000 by
PALGRAVE™
175 Fifth Avenue, New York, N.Y. 10010 and
Houndmills, Basingstoke, Hampshire, England RG21 6XS.
Companies and representatives throughout the world.

PALGRAVE™ is the new global publishing imprint of St. Martin's Press
LLC Scholarly and Reference Division and Palgrave Publishers Ltd
(formerly Macmillan Press Ltd).

ISBN 978-0-312-21712-9 hardback

Library of Congress Cataloging-in-Publication Data
Ranft, Patricia.
A woman's way : the forgotten history of women spiritual directors / by
Patricia Ranft.
     p.   cm.
   Includes bibliographical references and index.
   ISBN 978-0-312-21712-9
   1. Spiritual direction—History.   2. Women spiritual directors—
History.   3. Spiritual directors—History.   I. Title.
BV5053.R36   2000
253.5'3'082—dc21

                                                              99–059605

A catalogue record for this book is available from the British Library.

Design by Letra Libre, Inc.

First edition: November 2000
10   9   8   7   6   5   4   3   2   1

To Michael
For so very, very much

Contents

Acknowledgments

I would like to thank those friends whose conversations helped nourish me while working on this book. In particular, I am grateful to my dear friend Terry Fleming for allowing me to use him as a sounding board time and again. His reactions and corrections were always helpful. I am also indebted to Angela McGlynn for discussing logotherapy with me.

The rest of my thanks are due to the faithful few who came to my aid once again to fulfill tedious tasks. As usual, Annette Davis saved me from many a computer snafu and helped produce the final manuscript. My daughter Meredith and my husband Michael edited and proofread the manuscript, while my son Jeff and my son-in-law Oddur offered their constant support. I am most indebted to Michael for his very constructive criticism and, more importantly, for making me believe all things are possible. Finally, I would like to thank Meredith, Oddur, Jeff, Debbie, my parents, our siblings and their families, and all our friends who have offered much-needed support during the last few months; I am especially grateful to Robert and Kate for their help and concern. Michael and I are very fortunate indeed.

Introduction

Women's history has certainly come a long way in a very short time. Perhaps the reason for this phenomenal progress is the fact that so much of it was there, just waiting to be rescued from obscurity by interested historians. In many ways the progress parallels medieval history. At the turn of the nineteenth century, few historians dabbling in the Middle Ages could have predicted how vast our knowledge of this period would become; surely even fewer guessed that the period was as pivotal and crucial to the modern world as we now know it to be. Most amateur historians accepted the verdict of early modern humanists who wholeheartedly believed that the thousand-year period was a barren era of Gothic ("barbarian") culture in the middle of two glorious eras of human achievement. After a century of meticulous documentation, analysis, and research, we know that it was during the Middle Ages that much of what still shapes our world today was created, nourished, tested, and even sometimes perfected.

Women's history also began with limited hopes. Many scholars believed they would at best be able to construct a complementary history in which they could document peripheral roles women played in certain arenas that, while not determining the shape of history, did complete the historical narrative. Others feared that the end product would be subject to criticism of distortionism if they placed women's actions at center stage, for, they worried, had not historiography already determined that only male actors played starring roles? By focusing solely on women, the historically trivial might appear essential; the validity of the narrative might be thus negated. Still others argued that women's history would be nothing more than historical appendixes where women's actions were tagged on or listed after the "real history" was completed.

Few scholars anticipated the results of the first generation of women's historians. One of the reasons they were surprised is related to the general framework that dominates contemporary thought patterns: the subtle, pervasive belief in historical progress. Call it development, evolution, social Darwinism, progressivism, dialectic process, or whatever; most people in our culture possess a deep-seated belief that history is the story of all things eventually improving and advancing. If nineteenth- or twentieth-century women were not key agents in the history of their period, if they were perceived to be inferior to males in almost all realms, if they could not attain the chief goals of their society on an equal basis with men, then surely women were much worse off a thousand years ago. Because this belief in historical advancement is so pervasive in our modern culture, even when it is explicitly disavowed by participants in academic discourse, there is still a resistance to accepting the results of research. Increasingly historians are finding that women from antiquity and the Middle Ages possessed much more power in their society than previously thought—or than many possess in modern times. There are now three decades of research that document how women exerted their power by influencing people, achieving their goals, and acting effectively.[1] In part, the discovery of powerful women who shaped history was made possible by theorists analyzing the concept of power. The division of power into formal, public (read "male") power and informal, private (read "female") power, now commonplace, was a breakthrough for women's studies, for it provided historians with categories in which to assess women's impact on society. As with most theoretical approaches, there is now a second-generation debate among scholars that challenges the validity of such a definition and division of power. Regardless of the outcome of the argument, the approach served historians well at a crucial time in women's history.

Still, the fact that our study of women's power has been so prodigious should not stop us from seeking yet other ways to learn of women's past. Just like attachments to theories of historical progress can at times blind us to alternative historical interpretations, so too loyalty to the study of power may inhibit us from examining other significant aspects of society responsible for shaping history. That power and the powerful are a crucial and driving force in history is beyond debate, but such an admission is not the same as saying it is the only determinant. Perhaps because it dominates in our society we assume it dominates in all societies, but our values are not necessarily everyone's values. As historian Carolyn Walker Bynum so masterfully argues in her studies of late medieval society, simply because we are consumed with sex and money does not mean that medieval society was.[2] Yes, we can assert with interdisciplinary support that certain values and concerns are universal, but when we attempt to prioritize those concerns, the debate begins in earnest. Sex and money are preoccupations with all peoples, but so are law, order, religion, security, food, thought, status, control, and so on. The way a society chooses to prioritize these concerns gives it its distinctive identity. We also know that there are present in every society those who desire to determine priorities; they desire to become powerful. Many scholars maintain that this drive for power is indeed the chief motive in human life. If this is so, are we then back to focusing chiefly on power in women's history? Certainly I would agree that historians should continue to focus their attention on the exercise and type of women's power and on their roles in political history. However, not all agree that the desire for power is the chief motive in human life.[3] Freudians maintain that the will to pleasure is primary. If so, then that which provides the most pleasure in life holds the key to understanding human actions and should be the focus of historians. Contemporary historians' preoccupation with sex and money in the form of social history and economic history is in large part a response to this Freudian thesis.

There is a third school of thought, however, that of logotherapists. It is logotherapists' contention that the search for meaning is the primary motivational force within each human. According to psychiatrist Viktor Frankl, founder of this school, the will to power and to pleasure are secondary to the will to find meaning in life, and "self-actualization is possible only as a side effect of self-transcendence."[4] Without meaning, even total power and pleasure cannot ultimately satisfy. Recent studies on substance abusers, those who attempt suicide, the incapacitated, the paranoid, the imprisoned, and the depressed support many of the contentions of logotherapy: A life with power, with pleasure, but without meaning, is difficult to live.[5]

In fields other than psychology and psychiatry, scholars are coming to a new appreciation of the centrality of the will to meaning. Political theorists, for example, argue that work is a political right which the state must provide opportunities for, and philosophers and theologians attempt to justify this theory. It is in and through work that humans find a sense of meaning and purpose in their lives; Pope John Paul II's encyclical *Laborem exercens* is a profound theological reflection in defense of this political theory, a theory particularly favored by Marxists and socialists. John Paul II maintains that the opportunity to work is "a fundamental right of all human beings"[6] because "through work man *not only transforms nature,* adapting it to his own needs, but also *achieves fulfillment* as a human being and indeed, in a sense, becomes "more a human being.'"[7]

While historians have not taken an active, open role in this discussion, their writings in religious history have indirectly and perhaps unwittingly contributed evidence to the debate. For over a millennium religion in Western society was of paramount importance. Was it because religion fulfilled the will to pleasure? Certainly pleasure was derived in much of the religious culture the Christian West embraced. A walk through a medieval cathedral while a choir sings the "Salve, regina," and the sun streams through the magnificent rose window behind a raised monstrance provides enough evidence of the aesthetic pleasures religion sometimes provided. A glance at the *Summa theologica* or a penitential handbook, on the other hand, renders compelling proof that pleasure was a secondary, not primary, purpose of Christianity. Was Christianity embraced so enthusiastically because it gave people ordinarily excluded from power an avenue to pursue power? Here the evidence is more difficult to assess. Surely it was such an avenue for many. When we study the institutional history of Western Christianity, we find it replete with examples of how some people continually used religious structures to attain power within their society. When historians were preoccupied with institutional religious history, the intimate relationship between religion and power appeared proved. Once the focus changed to noninstitutional history, however, that conclusion seemed more vulnerable. On what basis does the historian argue that diarymaids and village smiths, and even local nobility spent their lives immersed in religion because it gave them power and pleasure, when in fact religion did not enhance their power or pleasure to any significant degree? If religion failed to fulfill their will to power and to pleasure, why did they continue to death in their pursuit of it? Why did their children continue in their footsteps? The alternative answer is, of course, that religion provided meaning for their life.

While logotherapists argue very convincingly that the search for meaning is the primary motivational force in man,[8] they do not deny the presence of secondary forces or that the attainment of meaning often results secondarily in the attainment of power and of pleasure. But they believe that neither power nor pleasure gives individuals the ability to endure the unendurable or to embrace death; only meaning provides that. Like Nietzsche, Frankl believes that "he who has a *why* to live for can bear almost any *how*;"[9] this was a lesson Frankl learned personally during his years as a concentration camp inmate. Logotherapists believe that meaning has the ability to transform a human experience, and herein lies the historical importance of the will to meaning. If humans can change a tragic experience into an uplifting one by an internal, autonomous act, then at all times the person retains a degree of freedom: "It is not freedom from conditions, but it *is* freedom to take a stand towards the conditions."[10] Excruciating suffering ceases to be a totally negative experience when the sufferer transforms his or her condition by giving pain the meaning of sacrifice, as in childbirth. An individual may have every freedom taken away from him or her except one: the freedom to determine one's own attitude toward and interpretation of an event. Even the most powerful tormentor cannot take away this freedom. Meaning and freedom, therefore, are forever bound together in the human condition. When historians document the growth of freedom within societies, they must remember to start at the beginning, with the freedom to bestow meaning. This is what we intend to do here. We will study a particular group of women in

Western society who exercised their freedom to bestow meaning to life and communicated that meaning to others.

Two more aspects of the search for meaning are noteworthy for the historians, its emphasis on responsibility and on the future. Since logotherapists are therapists and not philosophers, their goal is not to provide an abstract meaning of life but rather to help individuals respond to the specific meaning of their lives at a specific moment. The key to human fulfillment, then, is in the response, for the meaning of life "is not to be questioned but responded to, for we are responsible to life."[11] Each person is questioned by life, and each "can only answer to life by *answering for* his own life." In other words, logotherapy sees "in responsibleness the very essence of human existence."[12] Thus, to Frankl, "life ultimately means taking the responsibility to find the right answers to its problems and to fulfill the tasks which it constantly sets for each individual."[13] It also follows from these premises that the future is all-important. Instead of looking back or within, as the psychoanalytical approach directs, logotherapy focuses on the future, when the discovered meaning of life can properly and continually be responded to "not in words, but in acting, by doing."[14] The response to the meaning of life is manifest historically in the individual's actions and thus can be documented.

All this creates an interesting opportunity for historians of Western women. There is always danger in imposing theories in one discipline *in toto* into another, and I make no attempt to do so here. Still, the historian is foolish to ignore research in other disciplines capable of shedding light on the subject being investigated, particularly subjects that produce limited yields when approached in traditional ways. With proper adjustments and commonsense application, the insights psychologists offer us about the motivational force within humans can bear fruit in our quest for knowledge about women's role in history. All schools of psychological thought agree that the will to power, the will to meaning, and the will to pleasure, regardless of how they are prioritized, are present in every individual and are the motivational forces for human behavior. Currently we are making great strides in expanding our knowledge of Western women's exercise of power and their social contributions, and we must continue to do so. However, we can also expand our historical knowledge further by examining the contributions made by women for whom the driving force in their lives was the will to meaning. We will do this by focusing on women who found the meaning of life in the religious and then led others to the same conclusion. We are identifying these women as spiritual directors, for that indeed was what they were. These women believed that the reason humans were created was to find happiness in God; God provided the meaning of life, and they tried to direct others to the same conclusion. Every time a woman helped, advised, guided, inspired by example, instructed, preached, cajoled, reared, chided, or directed another to reach the same conclusion, she was in fact offering spiritual direction. She was a spiritual director. Such women directed others toward the spiritual realm because they believed that ultimately only the spiritual contained the meaning necessary to attain true happiness. We must remember that it was the women spiritual directors' understanding of life—their spirituality—which they were directing others to accept. They were, in fact, influencing the actions and decisions of others, and collectively these actions and decisions shaped society. We as historians can no longer afford to ignore this obviously influential position in society.

It may surprise many to discover that tracing the history of women spiritual directors is a relatively easy task, because women have been exercising the ministry of spiritual direction quite extensively throughout the whole of Christianity's history. Some might even object that such a history is nigh impossible since spiritual direction itself was not practiced until late in Christianity's history. Such an objection, though, is based on a common but very limited definition of spiritual direction that identifies a ministry as spiritual direction only if it resembles modern direction. It is a faulty approach, because, as logotherapists point out, the meaning of life must be specific to the individual. Effective spiritual direction must change its message to be relevant to the needs of the person being directed. Consequently, the history of direction is a history in which certain permanent traits are present in ever new and variable ways. Spiritual direction is always the act of directing someone to the spiritual, but the way that person is led changes with each situation. Contemporary spiritual direction tends to be heavily clerical and emphasizes such things as pastoral counseling, sensitivity training, prayer techniques, and soul friendship. Early modern spiritual direction, in contrast, concerned itself with progress in prayer life and its stages, while other periods directed in other ways: by example, by pithy dictums, by exhortation, by education, and even by simple, direct statements of faith. Once we free the term "spiritual direction" from our modern context, we can more easily identify the spiritual direction practiced throughout the centuries and those who offered it. What we find is an ample history of both spiritual direction and women spiritual directors. Women directors were present already in Jesus' ministry and simply directed others to Jesus himself, the source of all meaning. During the first millennium we find women directors guided many in unsophisticated, unreflective ways. In the high Middle Ages women spiritual directors became more self-conscious of their role and their methods, and by the late Middle Ages spiritual direction was fully recognized and promoted, particularly by women mystics. It is the early modern period, however, that popularized spiritual direction among both laity and religious, and it is during this period that many of the masterpieces of spiritual direction literature were written. As we approach the modern period we find spiritual direction present in a multitude of forms. Among new Protestant sects women spiritual directors adopted many of the elementary ways and methods of their medieval predecessors, while Roman Catholic women continued to institutionalize the position within religious orders in the figures of the novice mistress and the mother superior. Also, the availability of published works during the modern era provided wider access to spiritual direction and promoted a less personal method. One constant remains throughout all these variable approaches, though: the spiritual director who guided her directee toward life's purpose, happiness in God. Directees were drawn to women directors, because they believed the women had grasped how to attain the happiness the directees sought. Because these women helped others in this central, if not primary, drive within human nature, spiritual directors were a major force in society. We must begin to identify these women and analyze the content of their influential messages.

Some mention of the terms used in spiritual direction may be helpful before commencing this history. Given the thorough secularization our vocabulary has undergone in the twentieth century, often it is difficult to remember the reality represented by such pietistic words as spirituality, salvation, eternal life, sanctity, heaven, sin, and similar terms. They are words used to express the same ultimate realities that concern us today. From the very beginning Christians reflected upon the meaning of life and offered guidance to others on how to achieve meaning in life. Indeed, the chief message of Jesus was his revelation of life's meaning and his direction on how to enjoy it.[15] The followers of Jesus became the first Christians when they, as theologian George A. Maloney says, "began this process of Christian reflection by asking themselves what was the end meaning of their lives."[16] As Christians reflected, different kinds of spiritualities developed, for spirituality is simply the particular way in which one expresses the meaning of life after experiencing it. Spiritual direction followed, since it too is merely advice on how to live one's life in light of that meaning. Both spirituality and spiritual direction are personally unique, because, as logotherapists tell us, it "is not the meaning in life in general but rather the specific meaning of a person's life at a given moment" that satisfies the will to meaning.[17] So it is with spirituality. Each individual and each society must discover a way to apply the immutable beliefs of Christianity to its specific historical situation. The spiritual director uses the vocabulary of her specific milieu to express eternal realities. Our difficulty comes when we fail to translate the expressions of another era into the equivalent expression of our own time. To seek salvation[18] is to seek happiness; to be perfect, united with God, and made in the image and likeness of God is to fulfill one's human potential. Demonic forces and sin are barriers to happiness. Heaven is where the will to meaning is satisfied; and revelation is

God's communication of life's meaning. Most important, to Christians God is the source of all meaning and Jesus is the ultimate spiritual director guiding them to that source. When we immerse ourselves in the literature of spiritual direction we must remember this essential task of translation, lest we overlook the profundities of the direction and dismiss it as only piety. The spiritual director did not use the technical language of philosophers, psychologists, or political theorists, but the realities they discussed were the same, and the influence they exerted over the thoughts and actions of others was perhaps even greater. The person who provides others with a way to satiate the human will to meaning is indeed a powerful historical force to be reckoned with. It is time to rectify our neglect of women spiritual directors and their power, and we begin to do so with this brief survey of their history.

Chapter 1

Early Christianity

It would be hard to find someone to dispute the statement that all people desire to be happy. Beyond that, the disagreement begins. Thomas Aquinas tells us that "to desire happiness is nothing else than to desire that one's will be satisfied"[1]—but psychiatrists and psychologists cannot agree on what satisfies our will. Ancient philosophers also disagreed about the substance of happiness. Plato held that union with the One would bring happiness, while Epicureans argued that the balanced life brought happiness. Stoics said it was the acquisition of virtue, and so on.

Within Christianity, however, there are no disputes, because the answer is at the core of its beliefs. Happiness is "the prize to which God calls me—life on high in Christ Jesus" (Phil 3:14). This happiness is possible because "for our sakes God made him who did not know sin to be sin, so that in him we might become the very holiness of God" (2 Cor 5:21). Such happiness is the intended end for all humans "made in our image, after our likeness" (Gen 1:26), for "God has not destined us for wrath but for acquiring salvation through our Lord Jesus Christ" (1 Thes 5:9). Humanity's intended end, to be "perfect in holiness," is the end for the person "whole and entire, spirit, soul and body" (1 Thes 5:23). These principles are expressed differently throughout Christian tradition, but the essential message, whatever abstractions are used to express it, is simple. Made in the image and likeness of God, humans find happiness only when they become one with that Image. It is perhaps most succinctly and poetically stated by Augustine of Hippo: "You made us for yourself and our hearts find no peace until they rest in you."[2]

When we examine the New Testament, we find women were central in the quest for happiness and in the dissemination of Jesus' direction of how to attain it. For example, each of the five great women of John's gospel—Mary (mother of Jesus), Mary of Bethany, Mary Magdalene, the

Samaritan woman, and Martha—contribute significantly to this task. Mary was instrumental when Jesus performed his first public miracle at the wedding at Cana. Her response to her son's question (in which he pointedly used the word for *woman,* not *mother*[3]) "Woman, how does this concern of yours involve me?" is one of faith: "His mother instructed those waiting on table, 'Do whatever he tells you'" (Jn 2:4–5). Here we also see a woman directing those without faith on how to respond to Jesus. This is the first spiritual direction given during Jesus' public ministry and hence the first in Christianity.

In the long, detailed story of the encounter between Jesus and the Samaritan woman, we see another woman playing even a larger role in deciphering Jesus' meaning and in guiding others to accept her interpretation. First, John tells us the woman demanded an explanation for Jesus' approach to her: "You are a Jew. How can you ask me, a Samaritan and a woman, for a drink?" Jesus' reply to her explicit questioning about her sex firmly established the inclusion of women in Christianity's search for meaning. The only criterion for inclusion is this, Jesus answered: "If only you recognized God's gift . . . he would have given you living water." His explanation did not satisfy her, so "she challenged him" again. This time she understood his answer and its meaning was clear to her: "The water I give shall become a fountain within him, leaping up to provide eternal life." Jesus provides *soteria:* salvation, happiness, the satiation of the will. The woman responded to this knowledge accordingly: "Give me this water," she replied, because she saw in these words the key to understanding the goal of life and the means to attain it. "'I know there is a Messiah coming,'" she proclaimed, to which "Jesus replied, 'I who speak to you am he'" (Jn 4:4–42).

The woman's first actions after her own desire to understand was satisfied were to direct others to such an understanding. When the disciples, "surprised that Jesus was speaking with a woman," returned to the scene, the woman quickly "went off into the town" and urged the people to encounter Jesus for themselves. They accepted her guidance and "at that they set out from the town to meet him." While each person had to embrace the message individually, John is careful to emphasize that the origin of their belief was rooted in the woman's witness: "Many Samaritans from that town believed in him on the strength of the woman's testimony." In fact, this narrative establishes the Samaritan woman as Christianity's first evangelist. Once she bore witness and directed them toward the source of her understanding, the townspeople responded as she had: "They begged him to stay with them awhile," and as they listened personally to his message "Many more came to faith." The essential, irreplaceable role of the woman was emphatically stated in the townspeople's acknowledgment of her spiritual guidance: "As they told the woman, 'No longer does our faith depend on your story. We have heard for ourselves, and we know that this really is the Savior of the world'" (Jn 4:4–42).

There are several interesting aspects of this tale, above and beyond the very distinct role that a woman assumed during Jesus' lifetime as a spiritual director. The narrative reveals that a woman was the chief instrument of Jesus in converting the Samaritans and in laying the "basis for a universal affirmation of God's salvation."[4] The narrative also notes that Jesus accepted the title of Messiah from a woman and a non-Jew at a time when he was not accepting the title when spoken by a man or a Jew.[5] While we see the woman in only one encounter progress rapidly in her comprehension of Jesus—note the increasingly enlightened titles the woman uses when addressing Jesus: "sir" (Jn 4:11,15); "prophet" (Jn 4:19); "messiah" (Jn 4:25); "savior of the world" (Jn 4:42)—the male apostles in contrast remain confused (Jn 4:33), even after their long association with him. Last, when the apostles' misunderstanding offered Jesus the opportunity of clarifying his meaning, he talked at length about their missionary obligation to spread his message, and he reminded them that they were indebted to the woman for her work of spiritual direction among her people. "Others have done the labor, and you have come into their gain" (Jn 4:38).[6]

In his last journey to Jerusalem Jesus performed his most dramatic and greatest miracle, raising Lazarus from the dead. Consistent with the rest of John's gospel, women play a major role in the events. Although John repeatedly tells us that Jesus loved in the abstract, John mentions only four people by name whom Jesus affectionately loved in the flesh: two men, "the disciple whom he loved" standing next to his mother at the Cross (Jn 19:26) and Lazarus; and two women, Martha and Mary.[7] That two women were singled out is significant when we remember that love, according to John, provides the meaning to life and the reason for salvation: "God so loved the world that he gave his only Son that whoever believes in him may not die but may have eternal life . . . but that the world might be saved through him" (Jn 3:16–17).

We have with these two women two distinct approaches for the attainment of happiness. After Lazarus' death Martha immediately "went to meet" Jesus "when she heard he was coming" (Jn 11:20). Martha actively engaged Jesus when they meet, and she was rewarded with one of Scripture's deepest and most abstract explanations of the meaning of human life and death: "Jesus told her, 'I am the resurrection and the life: whoever believes in me, though he should die, will come to life; and whoever is alive and believes in me will never die'" (Jn 11:25–26). When Jesus asked her, "'Do you believe this?'" she responded with complete acceptance. "'Yes, Lord,' she replied. 'I have come to believe you are the Messiah, the Son of God'" (Jn 11:26–27). Meanwhile, "Mary sat at home" (Jn 11:20) until Martha told her she was being personally summoned by Jesus. As she approached him "she fell at his feet," and "when Jesus saw her weeping," he too was "moved by the deepest emotions" (Jn 11:32–33). Mary's more passive yet emotional response when faced with the hard realities of the human condition was not rebuffed by Jesus but was actually embraced by him. Jesus himself "began to weep" (Jn 11:35) and proceeded to pray to the Father for Lazarus' return to life, causing "many of the Jews who had come to visit Mary . . . to put their faith in him" (Jn 11:45).

While he did not spurn the manner in which Martha approached life, Jesus was clearly more favorably disposed to Mary's way. When Jesus returned to Bethany later to celebrate Lazarus' restoration, "they gave him a banquet at which Martha served" (John 12:2), a fact that is accepted without comment. When Mary, however, was criticized by Judas Iscariot for wasting money on "a pound of costly perfume made from genuine aromatic nard, with which she anointed Jesus' feet" (Jn 12:3), Jesus vigorously defended her, telling Judas to "leave her alone" (Jn 12:7). A similar version of this visit is told in Luke. Here the critic was Martha, and Jesus did make a value judgment concerning one's approach to life. Martha, "busy with all the details of hospitality," was not condemned by Jesus, but he did tell her that Mary, "who seated herself at the Lord's feet and listened to his words," had indeed "chosen the better portion" (Lk 10:38–42). The witness of these two women's distinct approaches to Christianity were then and remain so today the two chief models upon which spiritual direction organizes itself, one emphasizing the active life, the other emphasizing the contemplative.

Mary Magdalene, the last of John's major female figures, is to some degree difficult to treat historically, because during the Middle Ages her identity became blurred. Mary of Bethany, Mary Magdalene, and a public sinner who anointed Jesus' feet "when Jesus was dining at a Pharisee's" (Lk 7:37–39) all merged into one figure, along with some legendary aspects. The Mary Magdalene presented in the Resurrection narrative of John's gospel is, however, not confusing but overwhelmingly powerful. She was the first witness to the empty tomb, the first to see the risen Jesus, the first to receive Jesus' charge to bear witness to the Resurrection, and the first to tell the other disciples about the risen Jesus. These distinctions should indeed be fully contemplated and assessed. It is a woman's understanding of the meaning of the Resurrection that is at the origin of Christianity, and it is a woman's communication of that perception which guides the rest of the disciples to their fulfillment. Christian spirituality—indeed, Christianity itself—formally begins with Mary Magdalene.

The four gospels vary on some details, but all agree that Mary Magdalene and other women disciples were the central figures in the events surrounding the death and Resurrection of Jesus. All four gospels report that the women were the ones who "had followed Jesus when he was in Galilee and attended to his needs" (Mk 15:40–41)[8] and who stood by the cross until the end. While Peter was busy denying Jesus and Judas busy betraying him, Mary Magdalene and "many other women who had come up with him from Jerusalem" (Mk 15:41) were steadfastly "watching everything" (Lk 23:49). After Jesus was taken down from the cross and laid in a tomb, "the women who had come up with him from Galilee followed along behind" to see "how the body was buried" (Lk 23:55). After he was laid to rest "Mary Magdalene and the other Mary remained sitting there, facing the tomb" (Mt 27:61). We must remember that Mary Magdalene and the other women— Mary, the mother of James and Joseph; Mary, the mother of Jesus; Jesus' mother's sister; Mary, the wife of Clopas; and Salome, the mother of Zebedee's sons, are mentioned by name[9]—were open and loyal disciples of Jesus in the same atmosphere that made Joseph of Arimathea decide to be "a secret one for fear of the Jews" (Jn 19:38).

It is on the third day, however, that the witness of Mary Magdalene becomes the example par excellence for all Christians, including contemporary disciples. All four gospels tell us faith in the Resurrection begins with Mary Magdalene's experience. Again, this is a startling fact not emphasized frequently enough. It is a woman who first identifies the risen Jesus as Teacher (Jn 20:16). It is a woman who offers the first Christian spiritual guidance. It is a woman who is the first missionary. It is a woman who is the first person commissioned by "the angel of the Lord" to "go quickly and tell" others about the good news (Mt 28:2,7), even before the eleven are commissioned. In Matthew the angel's commission was reiterated by the resurrected Jesus himself, who appeared to Mary Magdalene "with the other Mary" (Mt 28:1) as "they hurried away from the tomb half-overjoyed, half-fearful, and ran to carry the good news to his disciples" (Mt 28:8). "Do not be afraid!" Jesus added (Mt 28:10), trying to calm and encourage the women, for in Mark when Mary Magdalene did "announce the good news to his followers," they all "refuse to believe it" (Mk 16:10–11). Luke tells us that Mary Magdalene was not alone in meeting disbelief. Mary Magdalene, Joanna, and Mary mother of James, along with "the other women with them also told the apostles, but the story seemed like nonsense and they refused to believe them" (Lk 24:10–11). Mary Magdalene and these women thus also become the first Christians ridiculed and rebuffed for their faith. Even when the men meet the resurrected Jesus on the road to Emmaus, and he painstakingly "interpreted for them every passage of Scripture which referred to him" (Lk 24:27), they still did not have the faith of the women, prompting Jesus to exclaim in frustration, "What little sense you have!" (Lk 24:25). The men only later came "to know him in the breaking of the bread" (Lk 24:35); the women needed no rituals enacted to comprehend the meaning of their experience and to communicate it.

Some biblical commentators deflect the pivotal role of Mary Magdalene in the Resurrection narratives by emphasizing two other passages. In Mark, the earliest of the synoptic gospels, after being told by the angel to spread the news of the Resurrection, Mary Magdalene, Mary, and Salome were "bewildered and trembling; and because of their great fear, they said nothing to anyone" (Mk 16:8). However, an anonymous, longer ending (Mk 16:9–20) was written in the first century and soon became attached to Mark's gospel. Today not all Christian traditions accept Mark 16:9–20 as canonical, but it was considered so for the vast majority of Christian history.[10] In this ending, "Jesus rose from the dead" and "first appeared to Mary Magdalene," who immediately "went to announce the good news" (Mk 16: 9–10). It reflects the events related in Luke and John, and its authenticity was unquestioned in the medieval legends that grew up around Mary Magdalene. Even if we end Mark with verse 8, however, Mary Magdalene still retains her dominant position in the Resurrection narrative. She still attended to his needs at the cross and followed Jesus' body to "where he had been laid" (Mk 15:47). She still "brought perfumed oils with which they intended to go and anoint Jesus" (Mk 16:1) and was the first one told the news of the Resurrection (Mk 16:6). Moreover, since we know that testimony to the Resurrection was eventually given, and that the followers of Jesus eventually believed, we know that Mark's ending is not the end of the story. No evidence eliminates the possibility of Mary Magdalene and the women being the witnesses; there is much evidence in Matthew, Luke, and John to believe that Mary Magdalene did in time "announce the good news to his followers" (Mk 16:10).

The second passage that challenges the centrality of women's involvement in spiritual direction is in John. As theologian Gary Burge observes, "in the fourth gospel, John always gains the upper hand,"[11] so it should come as no surprise that here it is John, not Mary Magdalene, the other women, or even Peter, who is the first to believe (Jn 20:8). If there is an attempt to make John and Peter the key actors in the dissemination of the meaning of Jesus' death, however, it fails. Mary Magdalene continues to dominate, for John tells us she alone (in John's gospel there are no women with her) was the first to go to the tomb, the first to realize "the Lord has been taken from the tomb" (Jn 20:2), and the first to spread the news. John tells us next that Peter and John ran to the tomb to see for themselves after Mary told them: "They were running side by side, but then the other disciples outran Peter and reached the tomb first" (Jn 20:4). Peter went in first and examined the evidence, but John, moments later, entered, "saw and believed." However, "with this, the disciples went back home," and the focus switches back to Mary Magdalene: "Meanwhile Mary stood weeping beside the tomb." Her persistence in the face of absurdity is rewarded, for even as she wept, she stooped "to peer inside" the tomb and saw two angels. "'Woman,' they addressed her, 'why are you weeping?'" No sooner did she finish explaining why than Jesus himself addressed her in the same manner: "'Woman, why are you weeping?'" The use of the title "woman" to address Mary Magdalene at this crucial moment is reminiscent of his use of the title at the wedding of Cana and at Jacob's well in Samaria. In each instance the woman plays a key role in furthering Jesus' revelation. Here Mary Magdalene realized the meaning of the Resurrection and responded by calling out "'Rabbouni!' (meaning Teacher)." Jesus' final words to her are a missionary mandate to "go to my brothers and tell them, 'I am ascending to my Father,'" to which Mary Magdalene obediently responded and "went to the disciples. 'I have seen the Lord!' she announced. Then she reported what he had said to her" (Jn 20:8–18). It is this remarkable sequence of events that earned Mary Magdalene her most popular medieval title, the apostle to the apostles.[12] It should also have earned her the title of spiritual director of spiritual directors, since it is by following her direction that the apostles begin their own journey into Christian spirituality.

As with so many other matters, John's gospel contains a richer and more profound description of women's contributions to the development of Christian spirituality. This is not meant to imply, though, that women are absent in the synoptic gospels. We have already noted the attention they give to Mary Magdalene. Women are very much present in all the synoptic gospels. There is no tale of the Samaritan woman in Matthew, Mark, or Luke, but they do have their own presentations of women, including some not present in John's gospel. Luke gives us "a certain prophetess, Anna by name" who offered the witness of a full life, first as wife for seven years, and then as a widow "constantly in the temple, worshipping day and night in fasting and prayer" (Lk 2:36–38). When Anna was eighty-four Jesus was presented at the temple and made manifest to Simeon and Anna. It was Simeon's role to pronounce a canticle and an oracle in recognition of "the Anointed of the Lord" (Lk 2:26); it was Anna's task to herald Jesus' arrival. Thus, "she gave thanks to God" for the opportunity to end her last years with a set purpose, by "talk[ing] about the child to all who looked forward to the deliverance of Jerusalem" (Lk 2:22–38).

Luke also relates the story of the public sinner who is often confused with Mary Magdalene. Here we see a woman determined to find happiness through love and service. After gaining entrance to a dinner for Jesus held in a Pharisee's home, the woman revealed her total commitment to love and service by washing the feet of Jesus with perfumed oil and her tears. The host did not understand the meaning of her actions, so Jesus explained. "You see this woman? I came to your home, and you provided me with no water for my feet. She washed my feet with her tears and wiped them with her hair. You gave me no kiss, but she has not ceased kissing my feet since I entered. You did not anoint my head with oil, but she anointed my feet with perfume. I tell you, that is why her many sins are forgiven—because of her great love. . . . Meanwhile, he said to the woman, 'Your faith has been your salvation. Now go in peace'" (Lk 7:44–50). The woman fully understood the message of Jesus and responded with meaningful behavior and faith. Complete happiness excludes all aspects of evil, so when Jesus forgave her sins because of her goodness, he was directing her toward the Perfect Good.[13] The example of her life, in turn, gave direction to those also seeking the elimination of evil and the attainment of the Good. Her faith in Jesus and his message saved her.

One of the more pervasive themes in the gospels is Jesus' insistence that faith makes one whole, that is, faith saves (from the Greek *sozo*, to make whole, heal, save),[14] and women frequently presented Jesus with the occasion to proclaim this. When the woman with a history of hemorrhaging believed that if she touched the cloak of Jesus she would be healed, Jesus responded to her touch by telling her "Your faith has restored you to health [*sozo*]" (Mt 9:20–22).[15] The Canaanite woman's faith likewise resulted in Jesus' enthusiastic admiration. Determined to have Jesus save her daughter from a demon, she pestered him to the point that the disciple entreated him to "'get rid of her.'" He tried to discourage her by telling her that his "mission is only to the lost sheep of the house of Israel." Instead of being dissuaded she "did him homage" and pleaded "that he not so restrict his saving mission. Jesus then relented and exclaimed, 'Woman, you have great faith!' And her daughter was made whole" (Mt 15:21–28).[16] Not only did the Canaanite woman exhibit exemplary faith, then, but also she presented a convincing theological argument against an exclusive messianic community. "That such a theological argument is placed in the mouth of a woman," comments theologian Elizabeth Schüssler Fiorenza, "is a sign of the historical leadership women had in opening up Jesus' movement and community to 'gentile sinners' (Gal 2:15)."[17] We have already seen how the Samaritan woman's faith contributed to the spread of Jesus' message to non-Jews; we see it again with the Canaanite woman. To both women happiness was meant for all, and they believed with unassailable faith that such happiness meant salvation through Jesus. In both instances the women exercised leadership not only in identifying the inclusive meaning of Jesus' salvation, but also in providing direction through the witness they personally bore.

Another theme in the gospels common to women is that of service. While there may be some hesitancy to highlight women's association with service there because of possible reinforcement of what many consider to be stereotyping of women as domestic servants, women are often presented in the gospels as servers; men are not. The conclusion that such presentations reinforce misogynist mores that limit women, however, is made invalid by a principle clearly articulated by Jesus himself in both Matthew and Mark: "Anyone among you who aspires to greatness must serve the rest, and whoever wants to rank first among you must serve the needs of all" (Mt 20:26–27 and Mk 10:43–44).[18] The correct conclusion is not, therefore, that the gospels reinforce misogynist stereotypes, but rather that they offer women as the most responsive to Jesus' directives and, therefore, the chief models to be imitated by all. We can even turn this on its head and argue that by not portraying men as servers, the gospels emphasize the need for men to turn to women for guidance. These conclusions are dramatically driven home by another aspect of the narratives: The women are described as servers to Jesus himself. The significance of this additional fact is made clear by yet another comment of Jesus' recorded in John: "'If anyone serve me, him the Father will honor" (Jn 12:26).

Women as servers abound in the gospels. Years before Mary introduced Jesus' saving mission at Cana, she declared her willingness to fulfill the task given her. "'I am the servant of the Lord. Let it be done to me as you say," she replied at the Annunciation (Lk 1:38). When she journeyed to visit Elizabeth, who was awaiting the birth of her son, Mary met her with the *Magnificat*, one of the most revered and repeated prayers of the New Testament. Again Mary referred to herself here as the Lord's "servant" (Lk 1:48). The Samaritan woman responded to Jesus' request that she serve him water by leaving him her water jar (Jn 4:28). In Luke, Martha diligently served Jesus during his visit and was intently "busy with all the details of hospitality" (Lk 10:38–42). After Lazarus' return from the dead, the people of Bethany "gave [Jesus] a banquet at which Martha served" (Jn 12:2). Mary served him at the same banquet by anointing Jesus' feet and drying them with her hair (Jn 12:3). The public sinner served Jesus in a similar manner. We are told by Matthew and Mark that throughout his public ministry in Galilee, Jesus had a group of women disciples who attended to his needs and who remained faithful to the end of the crucifixion.[19] Luke tells us more about these women. As Jesus journeyed through villages preaching the good news, the Twelve male disciples followed him "and also some women." Luke mentions Mary Magdalene, Susanna, a married woman Joanna, "and many others who were assisting them out of their means," clearly indicating that women served Jesus in financial ways as well as in physical ways (Lk 8:1–4). All the synoptic gospels include the story of how Peter's mother-in-law was restored to health by Jesus. With a touch of Jesus' hand, the fever she was suffering from left her "and she rose and served him" (Mt 8:15; Mk 1:31; Lk 4:39). Serving Jesus continued beyond his death for some women, for in Mark, Mary Magdalene, Mary the mother of James, and Salome "bought perfumed oils with which they intended to go and anoint Jesus" (Mk 16:1). In Luke immediately after the women were satisfied that Jesus' body was properly buried, "they went home to prepare spices and perfumes" with which to anoint the body on the first day of the week (Lk 23:56). In short, there is no dearth of evidence in the gospels of women fulfilling the task of service as mandated by Jesus. With their fulfillment of this obligation they came closer to perfection and thus to happiness. There was a purpose to their service: to have the Father honor them; therefore they could endure any hardship. They knew why. Their service and knowledge became the model future generations directed their own behavior toward.

The Acts of the Apostles and the epistles also contain evidence of women contributing the formation of the primitive community's unique spirituality and of their directing others to Christianity's message. First of all, Acts establishes the dominant presence of women in the very first act of the new community. After Jesus' command for all to bear witness, he ascended into heaven before the apostles' eyes (Acts 1:9), and they returned to Jerusalem to devote themselves to constant prayer. The author is quick to add: "There were some women in their company, and Mary the mother of Jesus, and his brothers" (Acts 1:14). These women were also present "when the day of Pentecost came," and " all were filled with the Holy Spirit" (Acts 2:1–4). Afterward, they devoted themselves to "the communal life, to the breaking of bread and the prayers," (Acts 2:42), thus "winning the approval of all the people." Because of such witness, "more and more believers, men and women in great numbers, were continually added to the Lord" (Acts 2:47). Women's witness as well as men's was guiding others to accept the message and meaning of Christianity.

Women were already models of Christian action before Peter arrived in Joppa. We are told of a convert named Tabitha who lived such an exemplary life "marked by constant good deeds and acts of charity" that her death brought tears to all the widows and led Peter to raise her from the dead (Acts 9:36–43). Saul's persecution of Christians in Jerusalem was explicitly described as applying to men and women. He "dragged men and women out, and threw them into jail" (Acts 8:3) and got letters to "empower him to arrest and bring to Jerusalem anyone he might find, man or woman, living according to the new way" (Acts 9:2).[20] In Philippi Luke and Paul retired to the bank of the river for "a place of prayer" and found a group of women already there. One of the women, Lydia, "already reverenced God, and the Lord opened her heart to accept what Paul was saying." After she believed she in turn convinced her family, and "she and her household" became Christians (Acts 16:11–15). In Thessalonica many Jews converted, as "did a great number of Greeks sympathetic to Judaism, and numerous prominent women" (Acts 17:4); in Beroea many "came to believe, as did numerous influential Greek women and men" (Acts 17:12). The emphasis on the women's social influence indicates Luke believed that their conversion would guide others toward the same end.

In the epistles we find numerous women fulfilling Jesus' mandate to serve. Evodia and Syntyche "have struggled at my side in promoting the Gospel," Paul writes to the Philippians, "along with Clement and the others who have labored with me" (Phil 4:2–3). In Romans he tells us of "a deaconess" (from the Greek *diakonos,* a servant, minister), who "has been of help to many"; of Mary, Tryphaena, and Tryphosa, who "labored long in the Lord's service" (Rm 16:6,12); and of Prisca and Aquila, who were Paul's "fellow workers in the service of Christ Jesus and even risked their lives for the sake of mine." Evidently Prisca and Aquila were well known spiritual leaders, for not only Paul "but all the churches of the Gentiles are grateful to them." They even had a "congregation that meets in their house" (Rm 16:3–5).

There is one final group of women mentioned in the New Testament who were in a position to guide people spiritually: prophets. We have already mentioned the widow Anna in the temple. Mary mother of Jesus and Elizabeth certainly prophesized. When Elizabeth heard Mary's greeting, she "was filled with the Holy Spirit and cried out in a loud voice: "Blest are you among women and blest is the fruit of your womb'" (Lk 1:41–42), to which Mary replied, "All ages to come shall call me blessed" (Lk 1:48). In Acts we meet the "four unmarried daughters gifted with prophecy" of Philip the Evangelist (Acts 21:9). Most significant is Paul's discussion in 1 Corinthians 11:4–5: "Any man who prays or prophesies with his head covered brings shame upon his head. Similarly, any woman who prays or prophesies with her head uncovered brings shame upon her head." While the main focus in commentaries is usually on the question of veiled women, what is often overlooked is Paul's acceptance, almost expectation, of women prophets. In both the Old Testament and the New the prophet always has been acknowledged as one to whom the community must listen to for direction. Thus, it is not surprising to see an Old Testament passage from Joel paraphrased in Acts: "I will pour out a portion of my spirit on all mankind: Your sons and daughters shall prophesy. . . . Yes, even on my servants and handmaids I will pour out a portion of my spirit in those days and they shall prophesy" (Acts 2:17–18; Joel 3:1).

Besides gleaning from Scripture what we can about nascent forms of spiritual direction and women, we also find the biblical principles upon which spiritual direction rests. One of the most dominate verses commented on throughout the history of Christianity is the Genesis definition of humanity as made "in our image, after our likeness" (Gen 1:26). The Hebraic metaphor was given a prominence within early Christianity beyond what it had within Judaism, a prominence that lasts into the contemporary world. *Imago Dei* is undeniably the center of early Christianity's understanding of humanity as well as the center of its dogmatic expression.[21] Paul uses it in key passages in Romans, 1 and 2 Corinthians, Colossians, and Hebrews, as do the authors of the earliest Christian documents such as the *Didache*, the *First Epistle of Clement*, the spurious *Epistle of Ignatius to Antiochians*, and the *Epistle of Barnabas*[22]; every second- and third-century author I have examined discusses the term *imago Dei* or refers to the verse. In the magisterial study *The Idea of Reform*, historian Gerhart Ladner meticulously documents how the idea of reform is "essentially Christian in its origin and early development," how "the idea of reform implies the conscious pursuit of ends," and how the Christian idea of reform is "reformation to the image of God."[23] What distinguishes Christian reformation from other renewal ideologies is the belief that the chief goal in life is to "re-form" oneself into the original *imago Dei* of the Garden of Eden. While the idea is primarily and in its original content concerned with the individual, "reform proved eminently well adapted to reach beyond the personal" renewal and thus quickly became "a facet also of the doctrine of the Church." In short, "the idea of the reform of man to the image and likeness of God became the inspiration of all reform movements in early and medieval Christianity."[24] We must direct each action in our lives, in other words, toward becoming the *imago Dei*, because only when we recover our original image we will find happiness and perfection.

Spiritual direction is that which helps us figure out how best to "reform" our lives, to become *imago Dei* once again and thus be happy. The New Testament provides fundamental principles for spiritual direction. It tells us that true knowledge of the image of God is attained when we know "the glory of Christ, the image of God" (2 Cor 4:4), while at the same time one "grows in knowledge as he is formed anew in the image of his Creator" (Col 3:10). After all, "eternal life is this: to know you, the only true God, and him when you have sent Jesus Christ" (Jn 17:3). Scripture tells us that the way to gain this knowledge is through following good example. The premier example is, of course, Jesus himself. The true discipleship for the Twelve began after they "left everything and became his followers" (Lk 5:11). In the thrice-repeated question of Jesus "Simon, son of John, do you love me?" the final word on the subject is Jesus' mandate: "Follow me" (Jn 21:19). The synoptic gospels record Jesus' reply to a request for spiritual direction: "'Teacher, what good must I do to possess everlasting life?'" asked a rich man. "'If you seek perfection, go, sell your possessions,'" Jesus replied. "'Afterward, come back and follow me'" (Mt 19:16–22).[25] The call to follow the example of Jesus is repeated numerous times by Paul, but with advice to follow the example of those who have already succeeded in being followers of Jesus. "Imitate me," Paul advises, "as I imitate Christ" (1 Cor 11:1). First and foremost "be imitators of God as his dear children," (Ep 5:1), and "take as your guide those who follow the example we set" (Phil 3:17). When people "imitate those who, through faith and patience, are inheriting the promises" (Heb 6:12), they in turn become "a model for all" (1 Thes 1:7). In this way all can learn how to "reform" themselves according to the image of God.

Paul clearly understands that any one who properly imitates Jesus, consciously or not, becomes an example for others to follow—and, when one provides an example for others to follow one is, de facto, providing spiritual guidance. No where in the literature of primitive Christianity do we find the phrases "spiritual direction" or "spiritual guidance." We do, though, find many people reflecting on the reality of spiritual direction, albeit using different terminology. Early Christians talked about their obligation to bear witness, and when we examine the phenomenon we find that as a result of bearing witness, others were indeed directed toward the Christian understanding of happiness. Just as Jesus' example directed people toward spiritual perfection, every person bearing witness was potentially a spiritual director.

A brief review of the scriptural mandate to witness may help clarify this thesis. The biblical theme of witness has a long, rich history, beginning with Yahweh's call to Israel to make known to other nations its experience of the divine. "Let all the nations gather together, let the people assemble! . . . You are my witnesses, says the Lord, my servants whom I have chosen to know and believe in me and understand that it is I" (Is 43:9–10). This witness was to encompass Israel's whole existence; everything it said and did was to reflect the message of Yahweh.

When John the Baptist transmitted the Old Law to the New, the obligation to witness was included, and John became the first to bear witness to Jesus. After being witnesses, "the Spirit descended like a dove" and rested on Jesus while John baptized. John exclaimed: "'Now I have seen for myself and have testified, 'This is God's Chosen One'" (Jn 1:34). Jesus himself discussed the importance of John's witness and of witness in general. It is essential in directing others toward happiness.

> If I witness on my behalf, you cannot verify my testimony; but there is another who is testifying on my behalf, and the testimony he renders me I know can be verified. You have sent to John, who has testified to the truth. (Not that I myself accept such human testimony—I refer to these things only for your salvation.) He was the lamp, set aflame and burning bright, and for a while you exulted willingly in his light. Yet I have testimony greater than John's, namely, the works the Father has given me to accomplish. These very works I perform testify on my behalf that the Father has sent me. Moreover, the Father who sent me has himself given testimony on my behalf. . . . Search the Scriptures in which you think you have eternal life—they also testify on my behalf. (Jn 5:31–39)

Without witness the meaning of Jesus' life would escape comprehension, for Christianity is first and foremost a historical religion with an incarnate God. Humans need the testimony of other humans in order to believe. Believers have the obligation, therefore, once the testimony of John and others had convinced them of Jesus' truth, to add their witness to the testimony. Nowhere is this mandate more plainly stated than in the very last command Jesus gave his followers before ascending into heaven: "'You are to be my witnesses in Jerusalem, throughout Judea and Samaria, yes, even to the ends of the earth.' No sooner had he said this than he was lifted up before their eyes in a cloud which took him from their sight" (Acts 1:8–9). That the disciples understood the seriousness of this final comprehensive mandate is manifest particularly in Acts, where the apostles refer to themselves in the post-Resurrectional world as witnesses more frequently than any other designation.[26] It is an inclusive title and one used to express the totality of their faith and their responsibility to guide others to faith by their testimony.

Finally, when the biblical references to witness are examined in their context, it is evident that the testimony is not limited simply to verbal communication. The obligation to bear witness in society establishes a relationship between the individual and society. The whole person is involved in that relationship, and, therefore, one's actions as well as words are capable of communicating testimony. "No one begotten of God acts sinfully," John tells us. "No one whose actions are unholy belongs to God nor anyone who fails to love his brother. This, remember, is the message you heard from the beginning: we should love one another" (1 Jn 3:9–11). And how did Jesus communicate this message? "The way we came to understand love was that he laid down his life for us" (1 Jn 3:16): actions, not words. Taken as a whole, one's actions in life become a testimony to the spiritual realities one believes in.

As early as the end of the first century, Christians understood that the most effective and most accessible form of witness was the witness of personal example. The *First Epistle of Clement of Rome,* for instance, acknowledges the power of personal witness time and again. Written to address the division tearing the church apart in Corinth, the author calls upon the community to reestablish harmony by contemplating the witness of "ancient examples" and "the most recent spiritual heroes."[27] This refrain is repeated throughout. "Let us stedfastly contemplate those who have perfectly ministered to His excellent glory"[28]; "let us be imitators also of those who in goatskins and sheepskins went about proclaiming the coming of Christ"[29]; "let us turn to every age that has passed"[30];"having so many great and glorious examples set before us, let us turn again to the practice of that peace which from the beginning was the mark set before us, and let us look stedfastly to the Father and Creator"[31]; and so on.[32]

Significantly, even at this earliest stage the first Christians continued the tradition emanating from Jesus' ministry and acknowledged women as witnesses directing others to their spiritual end. "To these men who spent their lives in the practice of holiness, there is to be added a great multitude of the elect," the author of *Clement* says, such as the women Danaids and Dircae, who "furnished us with a most excellent example" and "finished the course of their faith with stedfastness."[33] When discussing the witness of Old Testament figures, *Clement* emphasizes only Abraham's example as much as the harlot Rahab's. The witness of Judith and Esther is also offered as an example to follow.[34] Thus, the first Christians recognized, as Jesus himself did, the ability of women to be spiritual directors.

Personal example is, in summary, the chief method of spiritual direction advocated in *Clement*. Early Christians did not identify their witnessing specifically as a way to direct people spiritually, but then again, they did not even identify their witness as witness. Reflection, identification, analysis, and organization were tasks for other ages; early Christians simply *did*. Inherent in their direction through the witness of personal example was buried a principle that is extremely important in the history of spiritual direction: the principle of adaptability. Because witness and direction are primarily acts of communication, the articulator must make sure his or her message is spoken in a language that can be comprehended by the audience. Whether it be actions or words, the audience must be able to decipher the meaning if the message is to be received. Because the message must be in the language of the society, every time the language changes, the witness and direction must change also. Witness and direction, therefore, if they be effective must adapt to the audience. The message remains the same, but the method, means, and presentation of spiritual direction must vary as much as society does. We have seen that already in Jesus' lifetime spiritual direction had various forms. In its most basic form, Christian spiritual direction simply directed others toward the person of Jesus. Some spiritual directors guided by service and by example, others by the proclamation of Jesus' message. As we pursue the history of spiritual direction, we will continue to see numerous variations of methods and approaches, but we will also find some constants. One is that all spiritual direction attempts to guide people toward happiness; another is that during the entire history of Christianity, women were spiritual directors. Moreover, society, following the example set in Scripture, also accepted, supported, and followed these women.

Chapter 2

*Women Spiritual Directors
of Church Fathers*

Given the scarcity of documents written by women during late antiquity and the early medieval period, only rarely do we have the opportunity to hear a woman's voice discussing spiritual matters and direction. Nevertheless, we can establish beyond doubt that women were spiritual directors who formed their own spirituality and who directed others according to that spirituality. If all humanity is made in the image of God (Gen 1:26), if there be neither male nor female before God (Gal 3:28), and if all people are called to "the very holiness of God" (2 Cor 5:21), then there is nothing extraordinary about a woman guiding a man to salvation—a conclusion demonstrated in the story of Jesus and the Samaritan woman. The premise that women are men's spiritual equals is accepted and endorsed in all early Christian sources without exception.[1] Once a society acknowledges a belief in women's spiritual equality then the construction of women's own spirituality and their exercise of spiritual direction no longer appears surprising. Indeed, they follow logically. As we turn our attention next to late antiquity to identify its women spiritual directors and to see how that society reacted to them, we must be patient. Because there are no extant sources written by women directors during this period, we must rely entirely on the women's male directees. Fortunately, these sources are abundant, for many of the most prolific and influential writers of the period had women spiritual directors and wrote at length about them and their direction. Still it is not the direct testimony of the women themselves; while we may end the review of the period frustrated by the lack of primary sources by women, we are at least able to document the practice of their ministry.

The family of Macrina the Elder is a good place to start our study of women's spiritual direction in the age of the church fathers, for it is unique in Christian history for its renowned holiness. Macrina the Elder's grandsons Basil the Great and Gregory of Nyssa, both church fathers, are perhaps the most well known, but they are not the sole reason why posterity has long revered the family; the reason is that the family nourished three generations of saints. Macrina the Elder, her daughter Emmelia, and Emmelia's children Macrina the Younger, Peter, Basil, and Gregory are all honored as saints. What is pertinent to us here is the fact that the family recognized the women to be the guides directing them all to their spiritual ends. "What clearer proof of our faith could there be than that we were brought up by our grandmother, a blessed woman," wrote Basil. "I am speaking of the illustrious Macrina, by whom we were taught the words of the most blessed Gregory [Thaumaturgus], which, having preserved until her time by uninterrupted tradition, she also guarded, and she formed and molded me, still a child, to the doctrines of piety."[2]

Like her mother, Emmelia took an active role in the spiritual formation of her children, particularly her firstborn, Macrina the Younger. "The education of the child [Macrina] was her mother's task; she did not, however, employ the usual worldly method of education," Gregory tells us in his vita of his sister, "but such parts of inspired Scripture as you would think were incomprehensible to young children were the subject of the girl's studies; in particular the Wisdom of Solomon, and those parts of it especially which have an ethical bearing. Nor was she ignorant of any part of the Psalter, but at stated times she recited every part of it." Indeed, "the Psalter was her constant companion, like a good fellow-traveler that never deserted her."[3]

Apparently Emmelia laid the foundation well, for it is Macrina the Younger who as an adult became the spiritual guide for the family. When Basil, for example, entered adulthood, Macrina continued his spiritual formation. After Basil "returned after his long period of education, already a practised rhetorician," Gregory wrote, "he was puffed up beyond measure with the pride of oratory and looked down on the local dignitaries, excelling in his own estimation all the men of leading and position. Nevertheless Macrina took him in hand, and with such speed did she draw him also toward the mark of philosophy that he forsook the glories of this world."[4] Macrina had an even stronger influence on the spiritual development of the youngest child in the family, Peter. She "took him soon after birth from the nurse's breast and educated him on a lofty system of training, practising him from infancy in holy studies" and eventually became "all things to the lad—father, teacher, tutor, mother, giver of all good advice." Peter learned well, and during his adult years it was he who helped Macrina administer the double monastery at Annesi; "always looking to his sister as the model of all good."[5]

When Emmelia's second oldest son, Naucratius, died, it was Macrina who "raised [Emmelia] up from the abyss of grief, and by her own steadfastness and imperturbability taught her mother's soul to be brave." By persuading her mother of the Christian meaning of death, Macrina "so sustained her mother by her arguments that she, too, rose superior to her sorrow."[6] In later life "when the cares of bringing up a family and the anxieties of their education and settling in life had come to an end,"[7] Macrina reversed roles with her mother and "became her mother's guide and led her on to this philosophic and spiritual manner of life."[8]

In both his *Life of St. Macrina* and his treatise *On the Soul and Resurrection*, Gregory of Nyssa gives us substantial information about the substance of Macrina's spirituality and the influence it had on others. First of all, Gregory explicitly stated that his purpose in writing Macrina's life was "to prevent such a life being unknown to our time and the record of a woman who raised herself by 'philosophy' to the greatest height of human virtue passing into the shades of useless oblivion."[9] He intentionally included in the vita only "what we had learned by personal experience"[10] and did "not think of it advisable to add" the tales of "sublime wonders" that others report, "for most men judge what is credible in the way of a tale by the measure of their own experience. But what exceeds the capacity of the hearer, men receive with insult and suspicion of falsehood."[11] In other words Gregory believed the personal witness of Macrina would effectively guide people to Christianity; he wrote her vita to make her witness better known. Gregory hinted that he learned even this, the power of witness, from Macrina. When visiting her on her deathbed they started reminiscing about their parents' childhood, but it soon turned into a counseling session "as I [Gregory] told my own trouble and all that I had been through." Her advice was to ponder the hardships of their parents' lives and learn from their example. "Will you not compare your position with that of your parents?"[12] Macrina suggested to Gregory, hoping thereby that he would gain the proper perspective of his own search for happiness. As the example of his parents' lives helped him, so he believed the example of Macrina would help others attain happiness.

Given Gregory's description of Macrina's death and funeral, we certainly can see that he was not the only benefactor of her spiritual direction. When the nuns "could no longer subdue their anguish in silence" and started crying, Gregory joined in. "Indeed, the cause for the maidens' weeping seemed to me just and reasonable," because "it seemed as if they had been torn away from their hope in God and the salvation of their souls, and so they cried and bewailed in this manner: 'The light of our eyes has gone out. The light that guided our souls has been taken away.'"[13] Her spiritual direction was not even limited to her family or monastery, but extended throughout the countryside. Gregory tells us there were innumerable disciples "who called on her as mother and nurse" and whose grief was "saddest of all." These were the people "whom she picked up, exposed by the roadside in the time of famine" and "nursed and reared them, and led them to the pure and stainless life."[14] Macrina had affected and guided so many people that the monastery was quickly overwhelmed with "the multitude of men and women that had flocked in from all the neighbouring country" to lament her death.[15]

Gregory provided a detailed enough portrait of Macrina for us to be able to reconstruct the content of her spirituality. We have seen that she valued the role of the intellect in human development, hence her promotion of education in her brothers' lives. Nevertheless, Macrina did all she could to direct those under her charge not to assume an elite position in society on the basis of their intellectual achievements. Prideful Basil, fresh from his intellectual accomplishments, was immediately humbled once under Macrina's guidance. Even more important, Macrina believed one did not hold oneself above the rest of the community because of social class based on material possessions or on occupation. After Basil renounced his intellectual elitism, he next renounced all his property and then employed himself in "this busy life where one toils with one's hands."[16] When Macrina "became her mother's guide and led her on to this philosophic and spiritual manner of life," Gregory outlined the stages of spiritual development Macrina directed Emmelia through. First, Emmelia was weaned "from all accustomed luxuries." Second, "Macrina drew her on to adopt her own standard of humility." Third, Macrina "induced her to live on a footing of equality with the staff of maids, so as to share with them in the same food, the same kind of bed, and in all the necessities of life, without any regard to differences of rank." Fourth, they "cast away all vain desires, honor and glory, all vanity, arrogance and the like." Last, thus freed, "nothing was left but the care of divine things and the unceasing round of prayer and endless hymnody, co-extensive with time itself, practised by night and day. So that to them this meant work."[17] The end result was a life on the borderline between human and spiritual nature, a nature "free from human weaknesses."[18] The women were limited by the human condition, but it was made easier to endure as their "philosophy continually grew purer with the discovery of new blessings." They still lived on earth even while "they walked in high in company with the powers of heaven,"[19] a position Gregory called living the angelic life.

Macrina believed work and abandonment of social status so integral to "the angelic life" that both are themes throughout the vita. When her intended husband died, she dedicated herself to her mother's service "and in all respects fulfilled the required services" of many maids, "even going so far as to prepare meals for her mother with her own hands." Macrina also took up the void created by her father's death in matters concerning the family's estates and shared her mother's toils in paying "taxes to three different governors."[20] Gregory commented that Macrina's "hands never ceased to work" and that "God secretly blessed the little seeds of her good works till they grew into a mighty fruit."[21] Gregory tells us Macrina's advice to her mother for spiritual progress. First, "Macrina persuaded her mother to give up her ordinary life and all showy style of living and the services of domestics to which she had been accustomed before, and bring her point of view down to that of the masses, and to share the life of the maids, treating all her slave girls and menials as if they were sisters and belonged to the same rank as herself."[22] Later, after Naucratius' death, Macrina openly became "her mother's guide" and persuaded her to intensify her asceticism by sharing life with the household slaves "in all the necessities of life, without any regard to differences of rank. Such was the manner of their life, so great the height of their philosophy, and so holy their conduct day and night, as to make verbal description inadequate."[23]

What did Macrina hope to achieve by directing her mother to live a life "on a footing equality with the staff of maids" and "without any regard to differences in rank"[24]? Macrina was not attacking society's class structure as an end in itself, for she was not primarily social crusader. No, she was more radical than that. She was challenging society's definition of happiness and the right of society to limit her understanding of happiness to society's definition without her free consent. Within the social structure of the day, Macrina's birth gave her many rights—wealth, privilege, freedom from toil—and many responsibilities—marriage, continuation of the family line, maintenance of estates. The spiritual life Macrina advocated negated all these rights and responsibilities. She desired the angelic life, the term Gregory frequently used in the vita. It was a common term of the period used to describe a life within a harmonious, spiritual community that equal persons who had no bodily ties with another, freely choose to join.[25] To live the angelic life was to refuse to accept society's definition of happiness. Such a refusal also meant one was free from the obligation of following the social conventions of the day. As historian Peter Brown argues, ascetics of the fourth century believed that they were "no longer permeable to the demands that society made upon it."[26] Thus, Macrina was directing her mother to strike at the heart of the social structure of the day by refusing to fulfill any of its demands not freely agreed to. The individual controlled her or his own destiny by choosing her or his own definition of happiness.

Macrina's devotion to Christianity's most radical symbol, the cross, indicates that she understood the extreme nature of her position. Apparently it was a rather private devotion that Gregory was unaware of until her death. On her deathbed Macrina began her last prayer by summarizing the Christian meaning of life. "Thou, O Lord has freed us from the fear of death.

Thou hast made the end of this life the beginning to us of true life." Such a radical statement is beyond logic; it must be believed with one's heart. To help in this belief God gave "a sign to those that fear Thee in the symbol of the Holy Cross, to destroy the adversary and save our life."[27] After reminding Christ that "I, too, was crucified with Thee," Macrina ended her prayer by sealing "her eyes and mouth and heart with the Cross."[28] Later that evening she drew her last breath after she brought her hand "to her face to make the Sign."[29]

The full extent of her devotion was only revealed later when Gregory and a woman named Vestiana, who had "chose[n] the great Macrina as protector and guardian of her widowhood,"[30] prepared Macrina's body for burial. As they viewed a scar on Macrina's breast, Vestiana told Gregory a story about her devotion to the cross. When a tumor had grown on Macrina's breast, she refused to have a doctor treat it. Instead, she prayed, made a mudpack, and then asked her mother to cure her by making a holy seal over the breast. "When the mother put her hand within her bosom, to make the sign of the cross on the part, the sign worked and the tumor disappeared."[31] As they continued to prepare the body Vestiana called out to Gregory thus:

> "See," she said, looking at me, "what sort of an ornament has hung on the saint's neck." As she spoke, she loosened the fastener behind, then stretched out her hand and showed us the representation of a cross of iron and a ring of the same material, both of which were fastened by a slender thread and rested continually on the heart. "Let us share the treasure," I said. "You have the phylactery of the cross, I will be content with inheriting the ring"—for the cross had been traced on the seal of this too. Looking at it, the lady said to me again—"you have made no mistake in choosing this treasure; for the ring is hollow in the hoop, and in it has been hidden a particle of the Cross of Life."[32]

About fifty years earlier than this incident the true cross had been discovered, and by 335 Cyril of Jerusalem tells us "the whole world has since been filled with pieces of the wood of the Cross"[33]; apparently Macrina had obtained a particle of the relic. Given the surprise of Vestiana and Gregory when they found the cross hung around Macrina's neck, we can assume that the modern-day custom of wearing a necklace cross was unknown in their day[34] and that Macrina's devotion was singular.

Macrina's devotion to the cross and the paradox it symbolized was, nevertheless, consistent with her attitudes toward both her understanding of happiness and the social conventions of the day. Just as the crucifixion reordered every aspect of society, so, too, did Macrina's demand for the freedom to choose challenge society's understanding of life. Through suffering Jesus offered life to all people, and through suffering Macrina discovered the answers to the human condition. Gregory described in detail his first meeting with Macrina during his visit.

> I found her already terribly afflicted with weakness. . . . She discussed the future life, as if inspired by the Holy Spirit, so that it seemed as if my soul were lifted by the help of her words away from mortal nature and placed within the heavenly sanctuary. And just as we learn in the story of Job that the saint was tormented in every part of his body with discharges owing to the corruption of his wounds, yet did not allow the pain to affect his reasoning power . . . similarly did I see in the case of this great woman. Fever was drying up her strength and driving her to death, yet she refreshed her body as if it were dew, and thus kept her mind unimpeded in the contemplation of heavenly things, in no way injured by her terrible weakness. And if my narrative were not extending to an unconscionable length I would tell everything in order, how she was uplifted as she discoursed to us on the nature of the soul and explained the reason of life in the flesh, and why man was made, and how he was mortal, and the origin of death and the nature of the journey from death to life again. In all of which she told her tale clearly and consecutively as if inspired by the power of the Holy Spirit.[35]

Fortunately for us, Gregory did extend his narrative to tell the details of their discussion in a separate treatise, *On the Soul and the Resurrection.* Until recently the full significance of Macrina's presence in this treatise was not fully contemplated. Throughout the centuries most scholars acknowledged that *On the Soul and the Resurrection* was the discussion referred to in Macrina's *Life* and that therefore Macrina and Gregory probably did have such a dialogue.[36] That Macrina actually held the opinions and formed the arguments that Gregory puts in her mouth was never really considered. The possibility was dismissed outright for two reasons: because only someone educated in philosophy could write a treatise of such obvious Platonic shape and because Macrina was not a recognized intellectual capable of teaching the brilliant Gregory of Nyssa. Macrina's contributions were not denied; they were simply not considered to be primary. Now that we are focusing on her and we are more aware of the contributions of women to the doctrinal development within Christianity, however, the possibility of her authorship appears more probable. Without the introduction of any new evidence, but merely by viewing the material before eliminating Macrina's contributions outright, we can gain a deeper appreciation of her influence on Christian spirituality and, consequently, on spiritual direction.

We have previously established that Macrina was the spiritual beacon guiding the lives and thoughts of her family and those in her monastery.[37] Second, throughout the vita Gregory used the word *"philosophy"*: "she raised herself by philosophy to the greatest height of human virtue"[38]; "also by her own life [Macrina] instructed her mother greatly, leading her to the same mark, that of philosophy"[39]; "Macrina took [Basil] in hand and with such speed did she draw him also toward the mark of philosophy"[40]; "the life of [Macrina] became her mother's guide and led her on to this philosophic and spiritual manner of life."[41] When earlier theologians like Justin Martyr declared "I am a philosopher,"[42] they employed the word "*philosopher*" almost as a synonym for "*Christian.*" According to historian A. M. Malingrey, philosophy connotes much more by the fourth century.[43] It describes one searching for the Platonic perfect life,[44] which, to most theologians, still meant the Christian life—but a Christian life that included rational considerations. Gregory of Nyssa marks this transition in theology, for with him "philosophy becomes the handmaiden of theology."[45] One did not leave rational thought behind when one became a Christian, but, to the contrary, used it with new energy to explain the mysteries one accepted by faith. Indeed, "it is our duty not to leave the arguments brought against us in any way unexamined."[46] The Christian's end in life "is one and one only: it is this: when the complete whole of our race shall have been perfected from the first man to the last,"[47] and this includes the perfection of the intellect. Given Gregory's appreciation of philosophy and the intrinsic role it plays in the Christian pursuit of perfection, it is hard to deny that his application of the word "*philosophy*" to Macrina was not deliberately intended to include her intellectual activity. We are told in her vita that Macrina possessed "natural powers" which "were shown in every study to which her parents' judgment directed her," and that her education was the special task of her mother.[48] We know that Macrina successfully directed her youngest brother Peter's education, being both teacher and tutor.[49] Gregory, in fact, addressed Macrina as the Teacher throughout his treatise on the soul. In summary, there is every indication that Macrina grew up well educated[50] and little evidence to suppose that she was incapable of engaging in an intellectual dialogue with Gregory or incapable of original thoughts on spirituality. If she formed and influenced her whole family's faith, why should we not accept Gregory's own insistence that she influenced the way he understood his faith?

We have already reviewed Gregory's description of the method of spiritual direction Macrina used with Emmelia. It was not her only approach, for Gregory also reveals in the vita that Macrina adjusted her approach to fit the specific needs of Gregory at a specific time. After an eight-year absence Gregory decided to visit her when their brother Basil died, because he was so upset by the loss. "So far was she from sharing in my affliction," however, that instead of joining him in his grief, she treated "the mention of the saint as an occasion for yet loftier philosophy." Then her spiritual direction began. First, "she discussed various subjects"; second, she inquired "into human affairs"; and third, "she discussed the future life," the issue that was troubling Gregory so much, and her guidance was so "inspired by the Holy Spirit" that Gregory felt "lifted by the help of her words away from mortal nature and placed within the heavenly sanctuary."[51] In *On the Soul* he repeated this description of her method, telling us her goal was "to correct with the curb of her reasonings the disorder of my soul."[52] To this end he begins a full exposition of one of humanity's most difficult dilemmas, the mind's need for rational certainty of the existence of the soul after death. Gregory knew what Christianity says about it, but in his grief his faith failed to sustain him. "Divine utterances seemed to me like mere commands compelling us to believe that the soul lasts forever; not, however, that we were led by them to this belief by any reasoning," he complained. "Hence our sorrow over the departed is all the more grievous; we do not exactly know whether this vivifying principle is anything by itself; where it is, or how it is; whether, in fact, it exists in any way at all anywhere."[53]

Macrina answered all these questions one by one, grounding all of them in the spiritual definition of humanity as made in the image and likeness of God. The soul, "a competent instructress," is "an essence created, and living, and intellectual,"[54] which "necessarily possesses a likeness to its prototype in every respect."[55] God's nature is spiritual and eternal, and in this the nature of the soul resembles God. Unfortunately, with Adam and "the arrival of evil human nature was diminished"; with the Resurrection we are "born again in our original splendour."[56] After this final resurrection is happiness: "Every one of the things which make up our conception of the good will come to take their place; incorruption, that is, and life, and honor, and grace, and glory, and everything else that we conjecture is to be seen in God and in His image, man as he was made."[57]

In other works Gregory expressed very distinct views on gender based on *imago Dei* theology. He wrote two treatises devoted to this theology: *On the Words "Let Us Make Man in Our Image and Likeness"* and *What Is the Christian Name and Profession?* In both he defines Christianity as the imitation of God and the Christian as *imago Dei* and discusses the ramifications of both at length. Another treatise, *On the Making of Man*, is a detailed description of his anthropology, again based on *imago Dei*. His consistent theses in all these treatises is, first, that "when Scripture says that God created man, *man* means *universal human nature* and *included the whole human race*"; second, that God made man in his image and likeness and is present "in equal measure to every member of the human race"; third, proof of God's likeness is "the fact that all men possess a mind" and "on this score there is no difference between the first man that ever was and the last that will ever be"[58]; and fourth, while in God there is no distinction of male or female, when he made humans he made them thus as "a means of increase."[59] Because of these fundamental principles of human solidarity, Gregory possessed a deep appreciation of women's spiritual worth. That, at least, would be the conclusion most would reach before considering the impact Macrina the Elder, Emmelia, and Macrina the Younger had on his life. This consideration would lead to a reversal of the conclusion: Because he possessed a deep, personal appreciation of women's spiritual worth, he was led to uphold fundamental principles of human solidarity. The difference between the two conclusions is the acknowledgment of the influence the women's spiritual direction had on Gregory.

If he were the sole figure we could turn to for proof, the argument would indeed be weak. Such is not the case, however, for the other two Cappadocian fathers, Basil the Great and Gregory Nazianzen, provide similar evidence. Both received spiritual guidance from women, and both believe that salvation transcends divisions of gender and class. We have already heard Gregory of Nyssa's testimony about how the spiritual direction Macrina gave Basil at that key moment of young adulthood radically affected his life. Another area where Macrina profoundly influenced Basil is in monasticism. Increasingly, scholars are acknowledging her involvement in the formation of so-called Basilian monasticism. The sources make the chronology clear. Emmelia, Macrina, Naucratius, Peter, and numerous servants were all living a life of asceticism at Annesi before Basil renounced worldly fame, and some ten years before he arrived at Annesi ca. 357. Emmelia disposed of most of her property and Macrina all of hers before Basil followed their example. Macrina had formed a closely knit ascetic group dedicated to continual prayer and hymnody.[60] Macrina and Peter resided over a "school of virtue" where "the men's department [was] presided over by Peter" and "the women's side" by Macrina.[61] Basil wrote very influential monastic rules that were followed by innumerable monastic communities throughout the centuries, but as historian Verna Harrison says, Macrina "appears to be the true founder of what is sometimes called 'Basilian' monasticism."[62]

We also have testimony from Gregory about the spiritual guidance Basil received as a child from Macrina the Elder, quoted at the beginning of this chapter.[63] In this passage we have an even more radical view of the role of women as spiritual directors; Macrina is the receiver of tradition from church father Gregory Thaumaturgus and the transmitter of it to Basil the Great, thus participating in what theologian Kallistos Ware calls the apostolic succession of saints. According to Ware, as the apostolic succession of bishops passes on the authenticity of orders, so to the apostolic succession of saints passes on the authenticity of spirituality and even doctrine.[64] If we accept this doctrine, women's spiritual direction becomes quite significant, because it is a well-known, respected, and readily accessible way for women to participate actively in the magisterium of the church and its shaping of the message of Christianity. Certainly this was consistent in principle with Basil's opinion of women's spiritual nature: "For the virtue of man and woman is one, since also the creation is of equal honour for both, and so the reward for both is the same. Listen to Genesis. 'God,' it says, 'created the human; in the image of God he created him; male and female he created them.' And the nature being one, their activities also are the same; and the work being equal, their reward also is the same."[65] There is no equivocation here, and similar statements can be found throughout his work,[66] particularly in his *Homily on the Martyr Julitta*. "I am from the same lump," Gregory recorded in Julitta's voice, "as men. We have been made according to the image of God, as they also are. By creation, the female, with the same honour as the male, has become capable of virtue. And for what are we of the same race as men in all things? For not only flesh was taken for the fashioning of women, but also bones from bones. So in solidity and tautness and ability to endure, we are equal to men, and this is owed by us to the Master."[67]

Gregory Nazianzen, the third of the great Cappadocian theologians, likewise had personal experience of women's spiritual direction from his mother Nonna and his sister Gorgonia. His eulogy at the funeral of his father Gregory the Elder was filled with higher praise for his mother than for his father. He attributed his father's spiritual condition to Nonna: "She who was given by God to my father became not only, as is less wonderful, his assistant, but even his leader, drawing him on by her influence in deed and word to the highest excellence; but not being ashamed, in regard to piety, even to offer herself as his teacher."[68] When first married Gregory the Elder was not a devoted Christian, and Nonna "could not brook this, the being half united to God, because of the estrangement of him who was a part of herself, and the failure to add to the bodily union, a close connection in the spirit." Her solution was to convert him, and "on this account, she fell before God night and day, entreating for the salvation of her head with many fastings and tears, assiduously devoting herself to her husband, and influencing him in many ways, by means of reproaches, admonitions, attentions, estrangements, and above all by her own character with its fervour for piety." She eventually succeeded, for she was persistent, and as "the drop of water constantly striking the rock was destined to hollow it," so, too, Gregory the Elder relented and followed her guidance to Christianity.[69]

After establishing Nonna as his father's spiritual guide, Gregory described the nature of her spirituality and its influence beyond his father to the whole household. Her goal in life was beauty, "but one kind of beauty, that of the soul, and the preservation, or the restoration as far as possible, of the Divine image." Her method of attainment was to reject the worldly goals of humanity and to redirect energies toward the spiritual. "The only genuine form of noble birth she reorganized is piety, and the knowledge of whence we are sprung and whither we are tending. The only safe and inviolable form of wealth is, she considered, to strip oneself of wealth for God and the poor, and especially for those of our own kin who are unfortunate."[70] As a result of her piety she became a model par excellence for all to imitate. Her spiritual goals dictated her behavior in every instance.

> What time or place for prayer ever escaped her? . . . Who paid such reverence to the hand and countenance of the priests? Or honoured all kinds of philosophy? Who reduced the flesh by more constant fast and vigil? Or stood like a pillar at the night long and daily psalmody? Who had a greater love for virginity, though patient of the marriage bond herself? Who was a better patron of the orphan and the widow? Who aided as much in the alleviation of the misfortunes of the mourner? These things, small as they are . . . are in my eyes most honourable, since they were the discoveries of her faith and the undertakings of her spiritual fervour.[71]

Here we have the portrait of a woman whose spirituality focused on relationships. To Nonna the center of each person was one's spiritual being, and thus no relationship was complete without a union of spirits. Until both Nonna and her husband shared the same understanding of ultimate realities, she felt the lack of total union, an "estrangement of him who was a part of herself." Nonna had a deep appreciation for the marriage bond and a sophisticated comprehension of how sex functions to bring about this bond, but she also believed the person was more than merely the physical. A person was body and soul, and unless she added "to the bodily union, a close connexion in the spirit," the relationship would be lacking. Moreover, there was difference in the bodies of male and female; there was equality in their souls. Their marriage was successful, Gregory tells us, because "in virtue they were quite equally matched."[72]

In his eulogy at his sister Gorgonia's death, Gregory described the marriage in even more detail. Nonna and Gregory the Elder were "united together with a bond of one honour, of one mind, of one soul, concerned as much with virtue and fellowship with God as with the flesh." Gregory the Elder is "the ornament of men," but Nonna is "not only the ornament but the pattern of virtue." The fact that Gregory the Elder converted and became a bishop, after all, "was the result of his wife's prayers and guidance." According to Gregory, his mother's spiritual direction was so extensive that it was "from her that [his father] learned his ideal of a good shepherd."[73] Gorgonia saw all this and duplicated their balanced relationship in her own marriage. Her goal "consisted in the preservation of the Image and the perfect likeness of the Archtype, which is produced by reason and virtue and

pure desire."[74] Gregory believed her particular gift was the ability to recon-
cile "the two divisions of life," the physical, worldly need for pleasure and
the inner, spiritual need for meaning. In this personified by the state of marriage and virginity, "she was able to avoid the disadvantages of each, and to
select and combine all that is best in both." By so balancing these desires she
"prove[d] that neither of them absolutely binds us to, or separates us from,
God or the world (so that the one from its own nature must be utterly
avoided, and the other altogether praised)."[75]

Rarely do we associate such a balanced approach to life with the spirituality of the period, and we cannot help but wonder whether this failure to
do so is due to our concentration on male spirituality. Fortunately, their contemporaries were more open-minded. Granted, all these reports are narrated
from male voices who were personally influenced by strong women spiritual
directors, but this merely reinforces our thesis that women exercised significant influence on the shape of their society and on certain men in their society who, in turn, were very influential. While many claim the writings are
too bias to be trusted as historical, given that the purpose of these writings
was to glorify the women and spread their fame, this purpose is actually
more of a reason to judge them as historically valid. The safest way to ensure
a society will graduate a person to the stature of greatness is to present those
aspects of the person which personify the fullest expression of society's highest values, not to present the person as one who flaunted its conventions to
the detriment of society. Herein lies the challenge of these men. They
wanted society to recognize the greatness of these women, yet much of what
they did historically seemed in fact to flaunt conventions. The solution was
to find a way to present exceptional behavior as acceptable; to this end the
male authors employed metaphors of masculinity to describe the women's
behavior. I disagree strongly with those who interpret these metaphors as attempts to remake women into men, or as thinly veiled beliefs that only men
could attain salvation or that a woman could gain happiness only if she became a man.[76] Rather, I agree with the conclusion of those who posit that
these metaphors are attempts, albeit clumsy to us, by men to affirm "the full
humanity of women they admire in the best way their androcentric culture
allows"; the metaphors are "a way of transcending culturally entrenched
misogyny, not a reaffirmation of it."[77] They recorded the central role women
had in their lives because, however risky it was to include this reality, it was
historically true; they employed metaphors of masculinity to make the historical influence of the women more believable and acceptable.

Gregory Nazianzen went to great length to communicate the historical ac-
curacy of his portrait of Gorgoria and to convince others that he was not dis-
torting history because he was "too greedy for her fair fame."[78] He insisted that
"in praising my sister, I shall pay honour to one of my own family; yet my

praise will not be false, because it is given to a relation, but, because it is true, will be worthy of commendation, and its truth is based not only upon its justice, but well-known facts."[79] When Gregory exclaimed, "O nature of woman overcoming that of man in the common struggle for salvation, and demonstrating that the distinction between male and female is one of body and not of soul!"[80] he is at the same time emphasizing the extraordinary nature of Gorgoria's achievements while trying to make them more ordinary by comparing them to male achievements. Christian theologians were proposing a radical innovation to society in late antiquity, the spiritual equality of women and men, and they did so cautiously so as not to be rejected. Christianity proclaimed in its worship, devotions, creed, and theology that there is neither male nor female before God, but only persons made in the image of God who attain the fullness of their being by "re-forming" themselves into this original image. When we examine the theology of the Cappadocian fathers we find, as Harrison argues, their writing "are surely a long way from the misogyny which is sometimes ascribed uncritically to all early Christians."[81]

The last of the great Greek fathers, John Chrysostom, also provides evidence of strong women directors in his world as do the Latin fathers Ambrose, Jerome, and Augustine. In John's case the woman was Olympias, a noble woman of great influence in Constantinople who was his closest friend, defender, and spiritual confidante during his long and troublesome public career as bishop and reformer of the capital city. When John was made bishop in 398, Olympias was already an established spiritual leader in Constantinople where, John wrote to her, "you are like a tower, a haven, and a wall of defense, speaking with the eloquence of example and through your sufferings instructing both sexes to strip readily for these contests."[82] She had been ordained a deaconess by John's predecessor Nectatius and led a group of patrician women in charitable works and spiritual discussions. After John and she formed their spiritual partnership, she established a thriving monastery for women, an orphanage, and a home for the sick. When John's reforms resulted in his exile, Olympias' influence and close spiritual relationship was so widely recognized that his enemies "contrived a diabolical machination against both" and "made her appear before the city perfect."[83] Condemned to exile like John, Olympias "turned over her flock" in the monastery to "her spiritual daughter" Marina and spent the rest of her life in Nicomedia, "maintaining her rule of life unchanged there."[84] Her generosity was phenomenal, "for no place, no country, no desert, no island, no distant setting, remained without a share in the benevolence of this famous woman." John even went so far as to unabashedly conclude that "quite simply, she distributed her alms over the entire inhabited world," bursting "the supreme limit in her almsgiving and her humility."[85] Her spirituality motivated physical contributions as well as spiritual. She was always busy "supplying the widows, raising the orphans, shielding the elderly, looking after the weak, having compassion on sinners, guiding the lost." Like Macrina and Emmelia, Olympias "called from slavery to freedom her myriad household servants, [and] proclaimed them to be of the same honor as her own nobility."[86] Twice Olympias is called an image of God, and three times we are told she was worthy of the title confessor,[87] because "engaging in much catechizing of unbelieving women and making provision for all the necessary things of life, she left a reputation for goodness throughout her whole life which is ever to be remembered."[88]

Among the Latin fathers we know the least about Ambrose's spiritual relationship with women. His sister Marcellina made a public "change of attire" to signify her profession as a nun (or "taking the veil," as it is still called) in 353,[89] shared much of his ecclesiastical and spiritual interests. "As I do not wish anything which takes place here in your absence to escape the knowledge of your holiness,"[90] Ambrose confided in Marcellina repeatedly. He worried about her, as she did him,[91] and he treated her like an intellectual and spiritual equal throughout his letters and in the treatise *Concerning virginity*, written at her request. Beyond the fact that Marcellina played a major role in Ambrose's spiritual life, we have no documentation with which we can identify to nature of that role.

Such is not the case with Jerome. Jerome's spiritual relationships with women were famous and plentiful, as is much of his refreshingly personal and open correspondence with these women. In many ways it is unique in ancient epistolary. Nearly a third of all his extant letters are to women, but this apparently is only a fraction, for Jerome informs us that "how many letters I have written to Paula and Eustochium, I do not know, for I write them daily."[92] The nature of many of the epistles, therefore, is confidential, relaxed, and almost akin to journal writing. When we read Jerome's correspondence in its entirety we find a highly opinionated, socially misogynist male who openly admired women's spirituality and willingly admitted his dependence on their spiritual guidance. He did not always succeed in containing his misogynism to the social arena, but he never allowed it to flourish to the point that he ignored their spiritual guidance.

The first known exposure of Jerome to women's spiritual guidance and influence came during his stay in Rome in 382. A powerful Roman matron Marcella had introduced a loosely knit monastic community at her Aventine estates a few decades earlier, and by the time of Jerome's arrival the group's spiritual prominence was well established. Ambrose's sister Marcellina was associated with the group, as were Albina, Fabiola, Asella, Sophronia, Melanie the Elder, Paula, and Paula's daughter Eustochium, all women responsible for spreading monasticism throughout Rome and the East. Marcella was the guiding force behind the movement, and it was her example that directed others to adopt her pioneering spirituality. "In those days no highborn lady at Rome had made profession of the monastic life, or had ventured—so strange and ignominious and degrading did it then seem— publicly to call herself a nun," Jerome recorded. "Nor was she ashamed to profess a life which she had thus learned to be pleasing to Christ. Many years after, her example was followed first by Sophronia and then by others. . . . My revered friend Paula was blessed with Marcella's friendship, and it was in Marcella's cell that Eustochium, that paragon of virgins, was gradually trained. Thus it is easy to see of what type the mistress was who found such pupils."[93]

Jerome soon became a regular visitor to the Aventine community, and before many months the group asked Jerome to lecture them on Scripture. Thus began one of the most important spiritual relationships in scriptural studies, for Jerome was prodded time and again by these women to trans- late and exegete biblical books.[94] Marcella "never came to see me that she did not ask me some question concerning them, nor would she at once ac- quiesce in my explanations but on the contrary would dispute them."[95] Such an inquisitive mind soon elevated her to the status of biblical author- ity, so much so that when Jerome finally left Rome, "in case of a dispute arising as to the testimony of scripture on any subject, recourse was had to her to settle it."[96] Jerome even privately admitted to Marcella her intellec- tual equality: "Ask me all the other questions you want when we are alone together. In that way if we happen to show our ignorance, the secret will have neither witness nor judge and will be entombed in a trusty ear."[97] Fabiola was even more demanding of him. After years of practicing asceti- cism and charity (Jerome said she was "the first person to found a hospital into which she might gather sufferers out of the streets") in Rome, Fabiola traveled to Jerusalem and "for a short time took advantage of [Jerome's] hospitality." Fabiola quizzed him endlessly on biblical passages during the visit. Once, "when she asked me the meaning and reason of each of these, I spoke doubtfully about some, dealt with others in a tone of assurance, and in several instances simply confessed my ignorance. Hereupon she began to press me harder still, expostulating with me as though it were a thing unal- lowable that I should be ignorant of what I did not know, yet at the same time affirming her own unworthiness to understand mysteries so deep. In a word I was ashamed to refuse her request and allowed her to exhort from me a promise that I would devote a special work to this subject for her use."[98] In such subtle and not so subtle ways, Marcella and Fabiola de- mands guided Jerome toward his spiritual fulfillment.

It was Paula, however who by virtue of their friendship provided Jerome with constant spiritual direction in his life. "Of all the ladies in Rome but one had power to subdue me, and that one was Paula," Jerome confessed.

"Our studies brought about constant intercourse, this soon ripened into intimacy, and this, in turn, produced mutual confidence."[99] While Paula (widowed at age thirty-one with five children) never moved to the Aventine estate, she did spend much time there, engaged in particular with biblical studies. The same council that brought Jerome to Rome in 382 also brought the venerated monastic advocate Bishop Epiphanius who stayed with Paula in her home. Paula's desire for the monastic life probably originated during discussions with Epiphanius. It was furthered by her association with the Aventine group, which by this time included Jerome. In 386 Paula left Rome with her daughter Eustochium for the East. Jerome, meanwhile, also decided to leave Rome, and the two friends agreed to meet in Antioch. From there they traveled through the Holy Land and Egypt together until autumn, when they settled in Bethlehem for the rest of their lives. Paula founded two monasteries there, one for herself and one for Jerome. In a very real physical sense, Jerome owed his monastic life to Paula's direction.

From this point onward the relationships with Paula and her family were the most important ones in Jerome's life. It is impossible to read the letters of Jerome without acknowledging this fact. When Paula's daughter Blaesilla died, Jerome grieved as if it were his own daughter. When he picked up his pen to offer words of consolation to Paula, his emotions hindered him. "As I think of her my eyes fill with tears, sobs impede my voice, and such is my emotion that my tongue cleaves to the roof of my mouth," Jerome lamented. "But what is this? I wish to check a mother's weeping, and I groan myself."[100] He wrote at great length to Laeta, Paula's daughter-in-law, about "how to bring up our dear Paula," named after her grandmother.[101] He tried to persuade Laeta "hand her over to Eustochium" in the Bethlehem monastery so that the child "will thus become her companion in holiness.'" To persuade her to do so he affectionately added, "Moreover, if you will only send Paula, I promise to be myself both a tutor and a foster father to her. Old as I am I will carry her on my shoulders and train her stammering lips."[102] Eustochium was a special favorite of Jerome's, and her friendship filled some of the void left in Jerome's heart and spirit after Paula's death. In Jerome's opinion Eustochium was the most admirable of all Paula's children, but he never forgot who guided Eustochium to such perfection: "I do not sever the daughter from the mother."[103] It was Paula who was the guiding light for her family, her monastery, and Jerome, and it was the example of Paula's life that directed all these people toward their spiritual goal.

Although Jerome was never one to shy away from rhetorical flourishes, we still cannot deny the depths of his sincerity as he eulogized after Paula's death: "If all the members of my body were to be converted into tongues, and if each of my limbs were to be gifted with a human voice, I could still do no justice to the virtues of the holy and venerable Paula."[104] In Jerome's opinion Paula possessed all the means necessary to achieve the ultimate goal, salvation. Her earthly accomplishments were impressive. She was a model wife and citizen who "won approval from all, from her husband first, then from her relatives, and lastly from the whole city."[105] She was a loving grandmother who felt such tremendous joy "when she heard her little granddaughter . . . falter out the words 'grandmother,'"[106] and a devoted mother whose "feelings overpowered her and her maternal instincts were too much" to bear at the death of her children.[107] Furthermore, she was a scholar whose linguistic skills surpassed Jerome, for while he with much toil and effort partially acquired the Hebrew tongue, Paula learned it so easily and "so well that she could chant the psalms in Hebrew."[108] But it was, of course, her virtuous conduct that Jerome believed brought her to her goal: "She sowed carnal things that she might reap spiritual things; she gave earthly things that she might receive heavenly things; she forewent things temporal that she might in their stead obtain things eternal."[109] A quick learner, she knew Scripture by heart and the history contained therein but considered it useful only if it provided meaning in her life. Therefore, she still preferred to search for the underlying spiritual meaning and made this search "the keystone of the spiritual building raised within her soul." To this end she asked Jerome to help her and Eustochium interpret the Bible. Realizing he did not have the depth of spiritual knowledge Paula was looking for, he instead told her what previous church fathers had taught. Like Marcella before her, soon the pupil became the Socratic teacher: "Wherever I stuck fast and honestly confessed myself at fault," Jerome confided, "she would force me by fresh questions to point out to her which of many different solutions seemed to me the most probable."[110]

Her detachment from material possessions and her unshakable belief that poverty helped one attain happiness were so total that they left Jerome frustrated, and he confronted Paula. "I was wrong, I admit; but when I saw her so profuse in giving, I reproved her" by quoting Scripture and composing numerous arguments. "With admirable modesty and brevity she overruled them all" with her clear logic. "I, if I beg, shall find many to give to [a beggar], he will die; and if this beggar does not obtain help from me, who by borrowing can give it to him, he will die; and if he dies, of whom will his soul be required?" Still, Jerome persisted in his arguments, "but she with a faith more glowing than mine clave to the Saviour"[111] until Jerome acknowledged Paula's insight into the spiritual life and submitted himself to her guidance. Jerome also believed Paula "was too determined, refusing to spare herself or to listen to advice"[112] in matters concerning the relationship between her physical discipline and the spiritual life. Jerome told us a humorous anecdote in reference to Paula's physical discipline, one that indicated how compelling her power for spiritual direction was. Once, when doctors tried to get her to take "a little light wine to accelerate her recovery" and she refused, Jerome "secretly appealed to the blessed Pope Epiphanius to admonish, nay even to compel her, to take the wine." It backfired, for "when Epiphanus left her chamber" and Jerome "asked him what he had accomplished, [he] replied, 'Only this[:] that old as I am I have been almost persuaded to drink no more wine.'"[113] In the end Jerome admitted defeat and realized her way had brought her to her goal: "She has finished her course, she has kept the faith, and now she enjoys the crown of righteousness."[114] He had only to follow her example to attain the same end.

Finally, we must note the spiritual direction Paula supplied to the monastic movement in Bethlehem. Jerome seemed most impressed with the flexibility with which Paula ruled her monastery, adopting her spiritual direction to fit the individual needs of each nun: "When a sister was backward in coming to the recitation of the psalms or shewed herself remiss in her work, Paula used to approach her in different ways. Was she quick-tempered? Paula coaxed her. Was she phlegmatic? Paula chid her, copying the example of the apostle who said: 'What will ye? Shall I come to you with a rod or in love and in the spirit of meekness?'"[115] Overall, Paula's method of guidance was simple. She guided the women "rather by her own modest example than by motives of fear,"[116] and this was a method that Jerome also adamantly believed in. His correspondence is replete with references to the power of example in directing people to salvation. Here we see the importance of his relationship with all these women. He was personally influenced by their example and therefore was drawn to acknowledge and even argue explicitly for a woman's role in spiritual direction. Women's example directed other women to spiritual perfection. Jerome wrote in a eulogy for Lea, a member of the Roman circle, that Lea "instructed her companions even more by example than by precept."[117] He specifically wrote a eulogy for Asella, Marcella's daughter, so that young girls "may guide themselves by her example, and may take her behavior as the pattern of a perfect life."[118] Not only should women be guided by her example, though; "let bishops look up to her."[119] In matters of spiritual direction, then, Jerome clearly believed all should follow the good example wherever it be found. "Would that men would imitate the laudable examples of women," Jerome exclaimed. If "with us Christians what is unlawful for women is equally unlawful for men, and as both serve the same God both are bound by the same obligations,"[120] and if "both have one task, so both have one reward,"[121] then why would not women who receive that reward be capable of directing others how to fulfill the task? Jerome saw no reason to suppose them anything but able.

It is nigh universally acknowledged that the last of the Latin fathers, Augustine, is the most influential. What is also known but not often properly assessed is the spiritual direction a woman provided in his spiritual life. Most analytical studies of Augustine's development concentrate almost exclusively on intellectual influences, whether they take the form of philosophy or theology. Yet Augustine is such an integrated person that he defied compartmentalization; his philosophy, his theology, and his spirituality are different expressions of a single vision of reality. Since he philosophized and theologized in order to communicate his understanding of life, however, I would argue that those that he turned to for spiritual direction were personally as important to his search for happiness in perfection as were those who influenced him philosophically and theologically. *The City of God* probably had a more far-reaching effect on the development of Western culture, but if the personal developments described in *Confessions* had not occurred, *The City of God* would not have been written. In other words, his spiritual experiences informed his total outlook on life and his expression of that understanding.

Because we are fortunate enough to have *Confessions,* we can easily identify the most influential person in his spiritual life. Monica, his mother, stands out above all others as the spiritual guide and anchor, indeed, as the determinative relationship in his life. She was so important that Augustine promised "I will omit not a word that my mind can bring to birth concerning your servant, my mother," even though "there are many things which I so not set down in this book, since I am pressed for time."[122] Of the nine narrative books in *Confessions,* almost the entirety of Book Nine is devoted to Monica, by far the most attention given to anyone. Consequently, we possess a rather vivid portrayal of a mother painted by a son—with interpretations typical of the relationship. According to Augustine, his mother is attached to him "far more than most mothers," had a "too jealous love for her son," was "alarmed and apprehensive" about him growing up, was "deeply anxious" when he got ill, was "unduly eager for me" to get an education, and would not rest content until he was settled in marriage.[123] In other words, she was like most mothers.

What makes Monica distinct in the annals of motherhood is the intensity with which she directed his spiritual well-being. It was her primary concern. "Words cannot describe," Augustine wrote, "how much greater was the anxiety she suffered for my spiritual birth than the physical pain she had endured in bringing me into the world."[124] In Augustine's opinion, his years of searching for meaning ended only because of Monica: "In her heart she brought me to birth in your eternal light."[125] He was rescued "from the depths of this darkness because my mother, your faithful servant, wept to you for me."[126] She wept for him and prayed for him, but also tried actively to guide him in the right direction. At one point Monica refused to eat at the same table as he, hoping to bring to his attention "the blasphemy of [Augustine's] false beliefs."[127] She asked a well-known bishop and Scripture scholar "to have a talk with me, so that he might refute my errors, drive the evil out of my mind, and replace it with good."[128] Monica encouraged his studies, because she thought they would "help me in my approach to you"[129]; she tried to limit Augustine's relationship with his non-Christian father, doing "all she could to see that you, my God, should be a Father to me rather than he."[130] She nurtured a deep friendship with Ambrose of Milan "because he could show me the way of salvation."[131] Once Monica even sailed in stormy weather to Milan to be with Augustine when he "was in grave danger because of my despair of discovering the truth."[132] But most of all, "she poured out her tears and her prayers all the more fervently, begging you to speed your help and give me light in my darkness."[133]

In the end Monica's prayers were answered, for how could they have been denied "when she asked not for gold or silver or any fleeting, short-lived favour, but that the soul of her son might be saved?"[134] It is noteworthy that when Augustine talked about her prayers being answered, he found the reason for their success in the context of her life. "But would you, O God of mercy, have despised the contrite and humble heart of that chaste and gentle widow, so ready to give alms, so full of humble reverence of your saints, who never let a day go by unless she had brought an offering to your altar, and never failed to come to your church twice every day, each morning and night, not to listen to empty tales and old wives' gossip, but so that she might hear the preaching of your word and you might listen to her prayers?"[135] Nor were her prayers for conversion only for Augustine; she was as persistent with her husband, and "in the end she won" his conversion also before his death.[136] Augustine clearly believed all who knew her were brought closer to salvation, for "those of them who knew her praised you, honoured you, and loved you in her, for they could feel your presence in her heart and her holy conversation gave rich proof of it."[137]

As with Augustine's conversion, Monica pursued all her spiritual goals on many levels, actively and passively, and those around her were influenced by her words, actions, and advice. When Ambrose was involved in an imperial conflict and the church of Milan was under threat while "faithful people used to keep watch in the church, ready to die," Monica "was there with them, taking a leading part in that anxious time of vigilance and living a life of constant prayer."[138] While following Augustine to Milan "over land and sea" she "put her heart into the crew," relying on "the sure faith she had in you."[139] While on her deathbed she discoursed "to some of my friends" about the ultimate questions of the human condition, life and death.[140] During his adult life "whenever she could, she used to act the part of the peacemaker between souls in conflict."[141] This made a particularly strong impression, for Augustine knew "from bitter experience" how needful people were of someone "to put an end to their quarrels by kind words. This was my mother's way, learned in the school of her heart, where you were her secret teacher."[142]

It is in one of the most powerful and earliest Christian descriptions of the mystical experience, however, that we can glimpse at the nature of the spirituality Monica shared with Augustine. The narrative also articulates well a goal of spiritual directors and directees: to reach an "understanding for which we [Monica and Augustine] had longed so much." The fact that it is Augustine's voice and not Monica's narrating the event actually enhances its value to us, because we see how successful her guidance was in spiritual matters. Augustine turned to Cicero for worldly eloquence, to Ambrose for doctrinal explanations, and to Simplicianus for guidance in worldly matters, but it was to Monica he turned to when probing life's ultimate mysteries. The narrative opens with Augustine describing an intimate scene similar to many other situations the two shared. "My mother and I were alone, leaning from a window which overlooked the garden in the courtyard of the house," trying to relax and "refresh ourselves before our sea-voyage." Their conversation "in the presence of Truth" was "serene and joyful" as they began "wondering what the eternal life of the saints would be like," that life which is so unknown. Together they probed, hoping "in some sense [to] reach an understanding of this great mystery." By reason they were led to conclude that no bodily pleasure "was worthy of comparison, or even of mention, beside the happiness of the life of the saints." Together they began to abandon reason and follow "the flame of love," which "raised us higher towards the eternal God." Their conversion continued as "higher still we climbed, thinking and speaking all the while" until together they left the physical world behind and entered into the mystical: "While we spoke of the eternal Wisdom, longing for it and straining for it with all the strength of our hearts, for one fleeting instant we reached out and touched it."[143]

After the experience ended, together they "returned to the sound of our own speech" and together reflected upon its meaning. In "that brief moment my mother and I reached out in thought and touched the eternal Wisdom"; they received mystical knowledge of the meaning of life. All things "have the same message to tell, if only we can hear it, and their message is this: We did not make ourselves, but he who abides forever made us."[144]

Suppose, we said, that after giving us this message and bidding us to listen to him who made them, they fell silent and he alone should speak to us, not through them but in his own voice . . . the voice of the one whom we love in all these created things; suppose that we heard him himself, with none of these things between ourselves and him, just as in that brief moment my mother and I had . . . ; suppose that this state were to continue and all other visions of things inferior were to be removed, so that this single vision entranced and absorbed the one who beheld it and enveloped him in inward joys in such a way that for him life was eternally the same as that instant of understanding

for which we had longed so much—would not this be what we are to understand by the words "come and share the joy of your Lord"?[145]

Monica had guided her son well, and together they learned the meaning of
life and experienced happiness.

Despite the scarcity of sources, then, we have established some important facts in the history of spiritual direction simply by examining the lives
and works of some of the Greek and Latin fathers. First, people were guiding and being guided toward spiritual ends even at this early stage of
Christianity. The spiritual direction offered then was not identified as
such, it did not involve the psychological analysis of later eras, and it did
not identify different stages, types, and methods of spiritual development,
as did many future periods. Directors simply offered guidance, most
through the witness of word and example, toward a spiritual end where
they believed happiness—salvation—would be found.

Second, women were among these early directors and were readily accepted by the Christian community as such. Since women's spiritual equality was not challenged during the patristic era, there was no basis to reject
them as spiritual directors. If they could achieve spiritual perfection, their
example could guide others to perfection too. Third, women's spiritual direction was accepted by the most respected members of the church intelligentsia. The influence these women directors had on the spiritual
development of men whose theological expressions shaped the doctrine of
the church is now documented; future analysis of Christian tradition and
doctrine must acknowledge the women's role and include their contributions. Last, in Gregory's *Life of Macrina* and Augustine's *Confessions* we have
the earliest detailed descriptions of nonmonastic spiritual direction. It is especially noteworthy that both are descriptions of women spiritual directors.

Chapter 3

Early Monasticism and the Early Medieval West

The formation of a new culture in the wake of a spent Roman civilization was challenging, to say the least, and has been analyzed endlessly ever since it commenced. As historian William Bark has argued, between A.D. 300 and 600 the West reached a turning point.[1] The newly arrived German people could form a new society out of the various forces present, or they and the remaining Romans could continue on the road of minimal survival. The decision to forge a new civilization was in no way inevitable, nor was its success assured. Indeed, the odds certainly were against any such undertaking succeeding. It did succeed, though, thanks to the decisions made during the erroneously labeled "Dark Ages." Instead of a period marked by inertia and ineptitude we now see one characterized by energy, creativity, originality, and flexibility. All scholars acknowledge the role of Christianity played in the process, and all agree the German peoples adopted Christianity as the vehicle through which their new culture would be formed and communicated.

In order for Christianity to retain such a position, it had to provide a rationale for this new society. It had to give the people, therefore, an explanation for why a new society was necessary and why it was worth working toward. It had to present a definition of life that was meaningful. Throughout the centuries intellectuals have often noted the failure of Roman culture to fulfill this essential definition. In a life with meaning, however, sacrifice was eagerly embraced because it brought one closer to life's goal. Early medieval people worked energetically to build a new environment because they believed that they were constructing something meaningful. While it is true that the chief task of a society's intelligentsia is to define life's meaning, this is not to say it is exclusively their domain. As the West reached this crucial period between 300 and 600, individual members of the Christian intelligentsia were not able to satisfy the needs of the now-large Christian population sufficiently. Aid came in the form of a new institution, monasticism. Its origin, function, and goal are complex and multifaceted and have been dealt with extensively elsewhere[2]; here we concern ourselves only with those aspects that pertain to its contribution to Christianity's understanding of life and the direction it gave to Christians who desired to attain that understanding.

One of the more significant facts to note about monasticism is that it is a lay movement. In the pivotal fourth century, while the church's structure became increasingly hierarchical and clerical, the birth of a lay movement outside that structure was a counterbalance.[3] The movement was accessible to the vast number of Christians excluded from the hierarchy, specifically women, the married, and children, and the spirituality developed within monasticism became the model for lay spirituality.[4] Despite appearing antisocial and negative, monasticism is the opposite. It is an institution whose purpose is to create a community where all people have an opportunity to attain the fullness of being, that is, happiness. To achieve this they paradoxically have to negate communities that fostered inequality. All individual barriers to complete personhood also have to be eliminated. Through trial and error men and women slowly constructed ways of living, eventually encapsulated in monastic rules that would help them reach their goal. In this way every thought they had, every action they took, had meaning. "Every art," the abba Moses said to John Cassian, promoter of monasticism in the West, "and every disciple has a particular objective, that is to say, a target and an end peculiarly its own. Someone keenly engaged in any one art calmly and freely endures every toil, danger and loss."[5] The farmer, for example, plows and clears because he knows it will mean an abundant harvest in the end. "So also with our profession. It too has its own objective and goal to which, not just tirelessly but in true joy, we devote all our labors. The hunger of fasts does not weary us. The tiredness from keeping vigil is a delight to us. The reading and the endless meditation on Scripture are never enough for us. The unfinished toil, the nakedness, the complete deprivation, the fear that goes with this enormous loneliness, do not frighten us off."[6] Once hardships were viewed from the proper and true perspective, they ceased to be negative experiences. "In the beginning," explained amma Syncletica, "there are a great many battles and a good deal of suffering for those who are advancing towards God and afterwards, ineffable joy. It is like those who wish to light a fire; at first they are choked by the smoke and cry, and by this means obtain what they seek (as it is said: 'Our God is a consuming fire' [Heb 12:24]): so we also must kindle the divine fire in ourselves through tears and hard work."[7]

As abba Germanus told John Cassian, this is all meaningful behavior willingly embraced in monasticism "whose goal, as the apostle says, is eternal life."[8] Eternal life is so worthwhile that "everything we do, our every objective, must be undertaken" consciously and specifically as a means to attain this goal. "Fasting, vigils, scriptural meditation, nakedness, and total deprivation do not constitute perfection but are a means to perfection." Moreover, "any diversion, however impressive, must be regarded as secondary, low-grade, and certainly dangerous," if it does not directly contribute to the attainment of eternal life.[9] Actions in this life are all the more significant because, as abba Moses commanded John, "Let everybody know this. He shall be assigned to the place and to the service to which he gave and devoted himself in this life and he can be sure that in eternity he will have as his lot the service and the companionship which he preferred in this life."[10]

Abba Moses also told John Cassian that while monasticism has "eternal life as its goal," the immediate objective is purity of heart. "Therefore, we must follow completely anything that can bring us to this objective, to this purity of heart," he continued, "for a mind which lacks an abiding sense of direction veers hither and yon."[11] Abba Moses was not alone in his belief. Aphraates, the most influential early Syriac monk, agreed.[12] Basil deemed purification "the essential exercise of monastic asceticism,"[13] while Athanasius called purification of the heart the Christian virtue.[14] To abba Moses purity of heart was "a heart that is perfect and truly pure, a heart kept free of all disturbance," particularly a heart "free of the harm of every dangerous passion."[15] The passions themselves are not a barrier to happiness, only when they are misused. "There is grief that is useful, and there is grief that is destructive. The first sort consists in weeping over one's own faults," amma Syncletica taught, "in order not to destroy one's purpose, and attach oneself to the perfect good. But there is also a grief that comes from the enemy, full of mockery, which some call accidie."[16] As difficult as it is to deal with passions, we must recognize the role they play in attaining purity of heart, for "it is only through many trials and temptations that we can obtain an inheritance in the kingdom of heaven,"[17] amma Theodora observed. "If, being a sinner, you undergo all these [temptations,] remind yourself of the punishment to come," added amma Syncletica, "and do not be discouraged here and now."[18] Still, the task is difficult and discernment is needed. At all times "we must direct our souls with discernment,"[19] she warned, for failure to do so will result in the loss of the objective.

It is quite easy to see that these men and women were on previously uncharted waters. Up to this point we have been discussing the concept of spiritual direction in a most literal way, as direction given to attain or understand a spiritual goal. Likewise, we have identified as spiritual directors those women known to direct or guide others to that goal. With the coming of the desert ascetics we begin a new phase in the history of spiritual direction. The desert abba and amma were intensely eager to recover the purity of heart necessary to comprehend the meaning of life. To that end they sought control over mind as well as body and soul. Thus, we must continue to "watch for the attacks of men that come from outside us," but at the same time we must observe our inner selves and "repel the interior onslaughts of our thoughts," according to amma Syncletica.[20] Observation and analysis of inner thoughts became one of the major preoccupation of the desert ascetics, and they observed and analyzed these thoughts with profound insight.[21] "It is good not to get angry, but if this should happen" let the angry dissipate before the day ends, counsels amma Syncletica, remembering always to "hate sickness but not the sick person."[22] Poverty, humility, obedience, and temperance, she continued, are necessary because these virtues act as safeguards against tendencies to be self-centered.[23] "As long as we are in the monastery" searching for true happiness "we must not seek our own will, nor follow our personal opinion,"[24] for control of the mind is ultimately more important than control of the body: "Obedience is prefer- able to asceticism. The one teaches pride, the other humility."[25] The answer to life's meaning ultimately is found in one's mind, even though subduing one's body helps free the mind to pursue the search without distractions. "Give the body discipline," amma Theodora proclaimed, "and you will see that the body is for him who made it."[26]

The desert preoccupation with inner thoughts and drives was unique for its time. To date no other movement had attempted to probe the depths of human nature's passions like the desert ascetics. The pioneering effort had two historical corollaries. These men and women developed a keen sense of obligation to share their insights into the human condition with any who sought such knowledge. It led, in other words, to the recognition in a more public and explicit way of the role of the spiritual director. As a result, the spiritual director did more than point people in the direction of Christian- ity's answer to life's mysteries; the director formed a relationship with a di- rectee and offered specific advice that addressed interior conditions which hindered the directee from identifying Christianity's answer. Second, be- cause the desert ascetics were so aware of the competing forces of good and evil within each person, they possessed a deeper consciousness for sin. This had a profound, long-lasting effect on the church's understanding of sin and the sacrament of penance. One result was that during the later medieval pe- riod the role of the spiritual director often was subsumed by the confessor.

The first corollary was evident almost immediately, and along with the as- cetics acceptance of their obligation to direct others came the identification of the qualities necessary for spiritual direction. According to amma Theodora, "a teacher ought to be a stranger to the desire for domination, vain-glory, and pride; one should not be able to fool him by flattery, nor blind him by gifts, nor conquer him by the stomach, nor dominate him by anger; but he should be patient, gentle, and humble as far as possible; he must be tested and without partisanship, full of concern, and a lover of souls."[27]

While finding a person resembling such an idealized model might seem impossible, the ascetics persevered. They knew that a person might be a good guide for one particular person and not for another. "If I prayed God that all men should approve of my conduct," commented amma Sarah, "I should find myself a penitent at the door of each one."[28] Not everyone has the ability to see the inner strength of another, be they director or directee. An angel had to appear to abba Piteroum and tell him about a near-perfect spiritual guide before he was able to recognize her as such. Piteroum traveled to her monastery, where she led a hidden life and was thought to be mentally unbalanced by her companions. When Piteroum met her he exclaimed to her community, "'You are the ones who are touched! This woman is spiritual mother'—so they called them spiritually—'to both you and me.'"[29] Syncletica argued that the most important quality for a spiritual director was, as Piteroum learned to acknowledge, his or her personal spiritual progress: "It is dangerous for anyone to teach who has not first been trained in the 'practical' life. For if someone who owns a ruined house receives guests there, he does them harm because of the dilapidation of his dwelling. It is the same in the case of someone who has not first built an interior dwelling; he causes loss to those who come. By words one may convert them to salvation, but by evil behaviour, one injures them."[30]

I hope that what has struck the reader by now is how visible women were in every aspect of the development of monastic spirituality and spiritual direction. It should also be noted that in the Alphabetical Collection, the ammas' sayings are listed in their normal alphabetical order, interspersed with those of the abbas. They are not attached at the end as an afterthought but are presented as integral to the collection. That their sayings are less numerous is obvious and may be disappointing to the researcher, but what is more significant is the breath and depth of what is preserved. As monastic scholar Irénée Hausherr wisely reminds us, "what counts is the presence of these women; this is a fact that has doctrinal value, based on a principle."[31] As Christianity was developing ways to search for life's meaning, women were there among the pioneers in manner fully consistent with the tradition set in the New Testament.

The desert ascetics never hesitated to accept a woman's spiritual insight, because the ascetics had no doubt that women were spiritually equal to men. It was, therefore, not surprising to find some women more advanced in perfection than men, for as John Chrysostom says, "the wrestlings of virtue do not depend upon age or bodily strength, but only on the spirit and the disposition. Thus women have been crowned victors, while men have been upset."[32] When the angel scolded abba Piteroum for thinking "so much of yourself for being pious," and challenged him to go "see someone more pious than yourself,"[33] he was not surprised because that person was a woman. He simply accepted the reality, as does Palladius, the narrator of the tale, without notice. When Evagrius Ponticus, probably the most influential and insightful articulator of desert psychology, was tormented by his passions, he turned to one best able to offer him spiritual direction through the crisis. That that person was a woman, Melanie the Elder, was of no consequence. Although doctors had first looked at him and his psychosomatic manifestations and "could find no treatment to cure him," Melanie correctly discerned that the problem was spiritual. She addressed him thus: "'Son, I am not pleased with your long sickness. Tell me what is on your mind, for your sickness is not beyond God's aid.' Then he confessed the whole story." She immediately saw that root of the problem, prescribed a remedy, and he "was well again in a matter of days."[34]

Palladius tells us of many instances in which women were spiritual leaders directing others. Besides directing Evagrius, Melanie the Elder had a spiritual relationship with Rufinus of Aquileia; together they "edified all their visitors and united the four hundred monks of the Pauline schism by persuading every heretic who denied the Holy Spirit and so brought them back to the Church."[35] On a trip to Rome "she met with a most holy and remarkable man, a Greek named Apronianus, whom she instructed and made a Christian"; in her own family she "lent moral support to her own granddaughter Melanie and to the latter's husband Pinianus as well, and she even taught her son's wife Albina."[36] In the town of Antinoe there were twelve women's monasteries directed by amma Talis, who guided the sixty women within solely by love.[37] When Athanasius was fleeing from the false charges of Arians, he went to an unidentified woman for refuge, telling her "God made it clear to me that I will be saved by no one but you."[38] Juliana of Caesarea did likewise for Origen.[39]

Melanie the Younger was a strong spiritual director in her own right. She guided her husband Pinianus into the ascetic life and remained his spiritual director throughout their life, making all decisions, spiritual and material, for them both. When Pinianus wavered on giving up the dress of nobility when he began his ascetic life, Melanie's biographer Gerontius reported her saying "Be persuaded by me as your spiritual mother and sister, and give up the Cilican clothes." Fully accepting of her position as guide, "straightway he obeyed her excellent advice, judging this to be advantageous for the salvation of them both."[40] Empress Eudocia "received her with every honor, as Melanie was a true spiritual mother."[41] Whenever Melanie "heard that someone was a heretic, even in name, and addressed him to make a change for the better, he was persuaded."[42] Sometimes her guidance took the form of informal teaching. When she was visiting Constantinople, for example, "many wives of senators and some of the men illustrious in learning came to our holy mother in order to investigate the orthodox faith with her. And she, who had the Holy Spirit indwelling, did not cease talking theology from dawn to dusk."[43] She even offered herself as spiritual guide to passing acquaintances and "persuaded many young men and women to stay clear of licentiousness and an impure manner of life. Those who she encountered, she taught with these words: 'The present life is brief, like a dream in every way. Why then do we corrupt our bodies that are temples of the Lord, as the apostle of God states? Why do we exchange the purity in which Christ teached [sic] us to live for momentary corruption and filthy pleasures?'"[44] When she experienced the mystical life and was "wounded by the divine love, she could not bear to live the same life any longer, but prepared herself to contend in even greater contests." Her decision to live as a hermit, however, was thwarted because her spiritual direction was in such demand "and for this reason everyone bothered her."[45] Eventually this demand led her to establish "a monastery of ninety virgins, more or less, whom she trained as a group" and to whom she gave spiritual guidance.

> Her whole concern was to teach the sisters in every way about spiritual works and virtues, so that they could present the virginity of their souls and the spotlessness of their bodies to their heavenly Bridegroom and Master, Christ. First she taught them it was necessary to stay vigilant during the night office, to oppose evil thoughts with sobriety, and not let their attention wander. . . . She would say, "Sisters, recall how the subjected stand before their mortal and worldly rulers with all fear and vigilance; so we, who stand before the fearsome and heavenly King, should perform our liturgy with much fear and trembling. . . . As for pure love to him and to each other, we are taught by the Holy Scriptures that we ought to guard it with all zeal, recognizing that without spiritual love all disciples and virtue is in vain.[46]

We must remember that, as monastic scholar Claude Peifer observes, the head of any ancient monastery was essentially a teacher whose authority was charismatic rather than hierarchical and whose defining function was the direction of members toward spiritual perfection.[47] Abbesses differed not at all from abbots in this matter. Among nuns, Hausherr concludes, "the 'mother' had the same prerogatives and obligations as the father among monks."[48] Every woman, therefore, who was the head of a monastery was de facto a spiritual director.

The phenomena of women spiritual directors and creators of new spirituality continued unabated throughout the early Middle Ages within the Eastern monastic tradition. Unfortunately, we have only a few sources written by women, so analysis of their spirituality and direction can be done only through male intermediary voices. Still, there is enough evidence in the sources to infer that women carried on the tradition and that men encouraged and benefited from that tradition. The seventh-century Syrian Martyrios was one such monk. He wrote not only about the benefits he personally received from a holy woman named Shirin but about how all the monastic abbots of the time "considered her as a blessed spiritual mother,"[49] how "monks and other strangers to the world who shared her reverence for our Lord used to come to visit her from all over the place, for they held her as a holy spiritual mother. They would gather from different places as children coming for lessons in sanctity with her," and thus would be guided "both by her words and by her actions."[50] Martyrios' mother, who was in fact also a spiritual guide for him, brought him frequently to visit Shirin, who instilled in the boy "all the greater ardor for the life of perfection as a result of seeing and hearing her. This ardor grew stronger every day, until my desire that originated from that source as it were consumed my youthful days."[51]

Writing to Abbess Euphrosyne Theodore the Studite, one of the most influential and popular monks in early medieval Eastern monasticism, congratulated her on her successful counseling of the nuns and enthusiastically encouraged her to continue. "We hear again and again of the excellent things you do, mostly while directing the sisters, keeping them united in one soul through charity, being vigilant in what regards God," he wrote. "Rejoice, therefore, good teacher and true mother according to God."[52] Theodore also offered Euphrosyne advice on how to best fulfill her role as abbess, and we see in the discussion what he believed the ideal relationship was between a female spiritual director and her directees. "Direct your attention to guiding the sisters," he began, and always remember "that all their eyes are on you as upon God; they view you as a reconciler between them and God. If you act this way, it is clear that on their part they must not aspire to do anything but what you, their teacher, desire, order and declare. Truly, they should act like true daughters toward the mother, like the members of the body towards the head."[53] Theodore's use of the body of Christ metaphor in this context is significant. Here women are not only included in the body of Christ, since "all of us are the body of Christ" and therefore "joined in one body, partakers and co-heirs,"[54] but women are also included as head.

The hagiographer of Abbess Irene of Chrysobolanton tells us more about the method used by spiritual directors. He reported Irene praying for discernment "to know what my sisters are doing in secret" so she could offer appropriate guidance. When her prayer was answered she "called everyone of her sisters to her, by name. She let [them] sit beside her; and then, very naturally, she led the conversation to hidden, secret things. She aroused the conscience by adroitly touching upon the movements of the soul and the sisters' inner thoughts. Thus she provoked a confession of faults and repentance, and had a complete correction promised to her."[55] Obviously the insights of the earlier desert ascetics were alive and well in Irene's monastery.

When we turn our attention to Western monasticism, which from its onset was almost exclusively cenobitic rather than eremitic,[56] the evidence of women spiritual direction is plentiful. The vitae of abbesses portray them as women who fulfilled the double function of the authoritative superior of the group and its spiritual director; in those centuries before the reestablishment of a legalistic mentality in the West, the latter function was the more important. Given the basic voluntary nature of the community and its explicit goal, if the monastic community—male or female—was unable to provide the individual with a way to reach this goal, then there was no raison d'être. Ultimately, the survival and success of a women's community depended on an abbess's ability to mold individuals into a spiritually cohesive group capable of attaining that goal.[57] A woman had to direct the members in these spiritual matters. If Christianity had not already had a tradition of women spiritual directors, it is difficult to see how women's monasticism would have thrived in the West during its early days. That hagiographers considered the formal, more institutionalized authority within a monastic community less important than spiritual authority is to be expected, given the goal of hagiographers and the social context within which they wrote, but the emphasis on spiritual power in hagiography does not de facto render its position false or even exaggerated. The evidence and the inherent logic of the situation speak forcibly for itself. Radegund, the most popular of all early medieval women, was both queen and abbess and, therefore, had an extraordinarily strong basis for institutional authority; yet although it made her more visible, her exercise of institutional power did not lead her to be celebrated in three major biographical works. We know more about Radegund because of her ability to draw people toward their own spiritual goals. That was what made her so significant to her contemporaries. "Though married to a terrestrial prince," wrote Venantius Fortunatus in one of three vitae of Radegund, "she was not separated for the celestial one and, the more secular power was bestowed upon her, the more humbly she bent her will— more than befitted her royal status."[58]

One of the vitae about Radegund is doubly helpful to us, because it was written by one of her directees, Baudonivia, and consequently it presents aspects of Radegund's spirituality that would not have been as readily available to outsiders like Venantius Fortunatus and Gregory of Tours (the third of the hagiographers). Not only was Baudonivia, writing as an eyewitness, "relating what I have heard and attesting to what I have seen,"[59] she was also writing as one directed personally by Radegund. "The preceding book" by Fortunatus "described many of the rigors of her abstinence and servitude" that were known publicly, "but her behavior as a pauper was so discreet that even the abbess suspected nothing,"[60] Baudonivia confided. A directee had access to a director as no outsider did, besides possessing the director's trust and confidentiality. Hence, we have in Baudonivia's vita revelation of a vision Radegund received and talked about "most secretly to two of her faithful ones and made them swear to tell no one while she lived."[61] Baudonivia was also able to observe some of the principles that informed Radegund's spiritual direction. "She never imposed a task on anyone that she had not done first herself" was one such principle; another was to pursue knowledge actively at all times. Every time she met someone new, "she would question him closely about his manner of serving the Lord. If she learned anything new from him that she was not accustomed to doing, she would immediately impose it first upon herself, and then she would teach her congregation with words what she had already shown them by her example."[62] Not all work rested on the shoulders of the spiritual director, though; the directee had responsibilities to meet. When meaning was not readily available to the women after reading Scripture, for example, "she would say, with careful attention to our souls, 'If you do not understand what is read, why don't you search for it diligently in the mirror of your souls?'" As guide, she was always there in the background, thus "she never ceased to preach on what the readings offered for the salvation of the soul."[63] Baudonivia also shares with us Radegund's perception of herself as spiritual director.

> She so loved her flock, which, in her deep desire for God, she had gathered in the Lord's name, that she no longer remembered that she had a family and a royal husband. So she would often say when she preached to us: "Daughter, I chose you. You are my light and my life. You are my rest and all my happiness, my new plantation. Work with me in this world that we may rejoice together in the world to come. With complete faith and hearts full of love, let us serve the Lord. Let us seek Him in awe and simplicity of heart so that we may say with confidence to Him: "Give, Lord, what you have promised, for we have done what you commanded."[64]

The forthrightness as well as the logic of the concluding demand lays bare the vigors and strength of Radegund's spirituality and her direction. There are no apologies for it or any sense that Radegund saw her direction as an encroachment on clerical ministries. There was only supreme confidence in its correctness. Baudonivia emphasized this by telling us that Radegund repeated this belief often. It is also apparent that Baudonivia fully agreed with Radegund's perception of spiritual directors as pastoral ministers, even if Baudonivia did not understand all its implications. "Frequently [Radegund] would say sweetly, in a sort of veiled figure of speech that none could understand: 'Anyone who has the care of souls must be sore afraid of universal praise.' But no matter how much she wanted to avoid it," Radegund could not escape her ability to direct people to happiness. "Thus, whenever the infirm invoked her, they would be healed of whatever illnesses imprisoned them."[65] Baudonivia reinforced Radegund's self-image by repeatedly employing pastoral images to describe the latter's activities. She called Radegund "this good shepherdess" who "would not leave her sheep in disarray."[66] Throughout the vita Baudonivia referred to the community as "her flock" whom Radegund protected "with the sign of the cross,"[67] and to whom she preached and taught with dedication. "However much we recall of her love," Radegund's teaching and goodness were lost to them at her death, Baudonivia lamented, but continued to pray to Radegund that God "grant that you may herd before you the sheep you once gathered. Following the steps of the Good Shepherd, may you bring your own flock to the Lord."[68] Not only did Radegund reform herself in the image and likeness of God, but she did so in the specific image of one responsible for the salvation of all.

The writings of another woman, Caesaria the Younger, give us further insight into spirituality and direction during Radegund's time. Caesaria the Younger was the successor to Caesaria of Arles, cofounder of the monastery of St. Jean with her brother Caesarius. "At the beginning of the foundation," Caesarius wrote a rule for his sister's group that the women were then to "determine by diligent experiment" whether the rule was helpful in attaining monasticism's goal and "in harmony with reason."[69] Radegund learned about the rule and wrote to Caesaria the Younger to request a copy. The latter responded with a letter of spiritual guidance for Radegund. One of the most striking characteristics of Caesaria the Younger's spirituality is its immersion and reliance on Scripture: "Holy and good and laudable is the rule you have chosen to live by; but there is no teaching greater or better or more precious than the reading of the Gospel."[70] Caesaria's letter is glued together with Scripture quotes as well as repeated admonitions to "listen intently when divine Scriptures are read"[71] and the like. She cautioned Radegund to "let none enter who do not know letters," for they would not be able to read Scripture, and to make sure "all must be bound to memorize the Psalter" and "strive to fulfill all that you read in the Gospel."[72] Caesaria was not advocating a mere external reliance on Scripture, however. She made it very clear that Scripture must be internalized if its true meaning is to be known and make an impact. "One who desires to serve religion must struggle with the whole soul, with all the strength of faith" to avoid the passions of the body. "Therefore you should always be reading or hearing divine Scriptures for they are the ornaments of the soul."[73]

Like a good desert amma Caesaria offered more than abstractions and generalizations. She addressed Radegund's specific situation. "I know you have abundant wealth," Caesaria advised; therefore, Radegund must "give as much as you can to the poor."[74] Since Radegund is "nobly born," she must learn to "rejoice more in religious humility" than in secular dignity. Once married, she must "disdain the fires of lust" in order to "arrive at the coolness of chastity," remembering that all these struggles "will not be gone from you even to the end of life, for you will be secure from what is past only remaining careful for the future." To do so Radegund must "always ponder whence you have been and where you will deserve to come."[75] If she stayed focused on the goal, it will be attained: "Think of nothing else; presume to speak of nothing else; nor do anything else. Stand peacefully through everything for His place is a place of peace."[76]

Bede also wrote of women abbesses who functioned as spiritual directors to both men and women, the most remarkable being Hilda of Whitby. In the latter part of the seventh century Hilda founded a double monastery at Whitby in Northumbria, whose success and fame were due almost entirely to her spiritual guidance. There she directed men and women in the "observance of righteousness, mercy, purity, and other virtues, but especially of peace and charity," following "the example of the primitive church" where all were equal. "So great was her prudence that not only ordinary folk, but kings and princes used to come and ask her advice in their difficulties and take it. Those under her direction were required to make a thorough study of Scripture and occupy themselves in good works, to such good effect that many were found fitted for Holy Orders and the service of God's altar. Five men from this monastery later became bishops—Bosa, Aetla, Oftfor, John, and Wilfrid—all of them men of outstanding merit and holiness."[77] Not since the days of Macrina do we know of a spiritual director, male or female, with such prominent directees.[78] Hilda's spiritual guidance also influenced many who lived by her monastery; "she also brought about the amendment and salvation of many living at a distance, who heard the inspiring story of her industry and goodness." She was "called Mother because of her wonderful devotion and grace" by all who knew her, and she never ceased "to instruct the flock committed to her, both privately and publicly," even during her long fatal illness. For seven years she used her illness to show others the way "to serve God obediently when in health, and to render thanks to him faithfully when in trouble or bodily weakness."[79]

We do not know the content or the method of Hilda's spiritual guidance, but we do know it was highly regarded by her contemporaries. She is universally acknowledged to be one of the major forces in early medieval England. Bede mentioned other influential abbesses of the period, although all dim in comparison to Hilda. Etheldrada, founder and abbess of Ely, "displayed the pattern of a heavenly life in word and deed" for all her nuns to follow,[80] as did Ethelberga, Earcongota, and Hildilid. Among the Merovingians we know of Bertilla who, even before she became abbess of Chelles, "resembled the spiritual mother." We also know she "fortified her daughters and others" by using "spiritual arts as precautions to lead them to good deeds," although we are not told what those spiritual arts were.[81] Sadalberga, abbess of Laon, followed the example of "the holy women Melanie and Paula" in her direction[82]; Gertrude ruled the monastery of Nivelles until her health interfered, thereupon "relinquishing all her temporal offices and the care she had been taking for her flock, except in spiritual"[83]; Rictrude, abbess of Marchiennes, first imitated Martha's external service, but later "she became Mary" and "changed the habit of her mind."[84] We have a bit more information about the spirituality of Aldegund, abbess of Maubeuge. A mystic, "the Holy Spirit emitted rays upon her like the sun and the moon shining through the inserted windows" of the "secret chamber of her house." These special graces taught her that "Our Lord deign to reveal the spiritual concerns of His salvation to our human gaze"[85] and led to her community "rejoicing with a single voice" at the privilege of being placed "under the spiritual yoke of this mother."[86]

A final example, that of Leoba, may be the most influential, for the missionary work of Boniface and Leoba played a highly significant role in the Christianization and subsequent unification of Germanic people into a cohesive medieval culture. When Rhabanus Maurus, the famed teacher and probably the greatest scholar of his age, asked Rudolf of Fulda to write Leoba's life, he hesitated because he had no first hand knowledge of the woman. Because of this Rudolf seemed to be a more factual hagiographer than most; he even provided a detailed description of his research methods and the measures he undertook to make sure his portrait of Leoba was historical. According to Rudolf's research, much of Leoba's power and reputation rested on her ability to direct others spiritually *verbo et exemplo,* a theme Rudolf expounded repeatedly throughout the vita. Rudolf's appreciation for the role of the spiritual guide is evident from the onset. Even as Leoba's spiritual direction is key to understanding her importance, so to is it necessary to know about Leoba's own spiritual director. "Before I begin the narration of her remarkable life and virtues, it may not be out of place if I mention a few of the many things I have heard about her spiritual mistress and mother, who first introduced her to the spiritual life,"[87] thereby providing us with yet another portrait of a woman spiritual director. Leoba's director was Abbess Tetta, who "ruled with consummate prudence and discretion" at Wimbourne, the monastery Leoba was trained at. She "gave instruction by deed rather than by words, and whenever she said that a certain course of action was harmful to the salvation of souls she showed by her own conduct that it was to be shunned."[88]

One particular sermon made an impression on Leoba, for she recalled it "with pleasure when she told her reminiscences." When an unpopular novice mistress died and those who disliked her uttered "bitter curses over her dead body to assuage their outraged feelings," Tetta knew the women needed some strong corrective guidance. "She therefore called all the sisters to gather and began to reproach them for their cruelty" and told them that one of the fundamental principles of Christian perfection is to "be peaceable with those who dislike peace." Then "she counselled them to lay aside their resentment." They must always remember that "if they wished their own sins to be forgiven by God they should forgive others from the bottom of their heart" and "forget any wrongs inflicted by the dead woman." Last, they should pray that God "would absolve her from her sins." Only after they all addressed the interior dispositions that led to the deplorable behavior and "agreed to follow her advice" did she then impose physical discipline. She "ordered them to fast for three days," so they could "to give themselves earnestly" to praying for the woman's soul without distraction.[89] Thus, in this one episode we can clearly see Tetta's approach in spiritual direction. First, identify the principle that is central to making one happy in a situation. Second, expose and control the passion that is interfering with happiness. Third, focus on the damage done by the negative passion and begin reparation. Fourth, eliminate all possible distractions, so the work of reparation will be completed successfully and happiness attained. Physical discipline plays only a secondary role in the process, that of a means to an end.

Rudolf's description of Leoba herself presents a woman eager to receive guidance wherever she could find it: "She learned from all and obeyed them all, and by imitating the good qualities of each one she modeled herself on the continence of one, the cheerfulness of another, copying here a sister's mildness, there a sister's patience." In one aspect, however, she was emphatic: her love of learning. Before she was even an adult "she had no interests other than the monastery and the pursuit of sacred knowledge." She obediently "worked with her hands at whatever was commanded her," but her preference was undeniable; "she spent more time in reading and listening to Sacred Scripture than she gave to manual labour."[90] She excelled in intellectual pursuits, and soon "Leoba's reputation for learning and holiness had spread far and wide." It was her combination of intellectual acumen and spiritual depth that led Boniface to ask her to help in his missionary work in Germany. He believed that through learning and holiness the Germans could be led to accept Christianity's view of life. He instinctively knew "that by her holiness and wisdom she would confer many benefits by her word and example." The emphasis on knowledge in the vita indicates the missionaries did not believe that the Germans would be converted to Christianity unless it answered the questions of the mind and presented a cohesive worldview to replace the pagan one. "In furtherance of his aims" Boniface founded a monastery at Bischofsheim and placed "Leoba as abbess over the nuns." There she was to direct them "according to her principles," and once again Rudolf emphasized her intellectual activities.[91]

> So great was her zeal for reading that she discontinued it only for prayer or for the refreshment of her body with food or sleep; the Scriptures were never out of her hands. For, since she had been trained from infancy in the rudiments of grammar and the study of the other liberal arts, she tried by constant reflection to attain a perfect knowledge of divine things so that through the combination of her reading with her quick intelligence by natural gifts and hard work, she became extremely learned. She read with attention all the books of the Old and New Testaments and learned by heart all the commandments of God. To these she added by way of completion the writings of the Church Fathers, the decrees of the Councils and whole of ecclesiastical law.[92]

To Leoba learning and praying had the same goal. Both helped a person find union with the ultimate reality. She never neglected one for the other but built her spirituality on a balance between the mind and soul, always being "deeply aware of the necessity for concentration of mind in prayer and study, and for this reason took care not to go to excess in either."[93] That such a spirituality would be helpful in converting a "people riddled with superstition and unbelief" is not surprising, and the success of Boniface and Leoba's missionary efforts in Germany is due in no small way to her spirituality and direction. Leoba "made such progress in her teaching" that "there was hardly a convent of nuns in that part which had not one of her disciples as abbess."[94] Her guidance was not limited to monks and nuns, either: "The princes loved her, the nobles received her, the bishops welcomed her with joy." Even Charlemagne "received her with every mark of respect," all because "of her wide knowledge of the Scriptures and her prudence in counsel."[95] In case any contemporaries doubted the role Leoba's spirituality and direction played in Boniface's own spirituality and in the conversion of the Germans, on his deathbed Boniface left instructions with all "senior monks of the monastery" to bury her next to him "so that they who had served God during their lifetime with equal sincerity and zeal should await together the day of resurrection."[96]

While Leoba's spirituality and direction is the most documented of the German missionaries, it is apparent in correspondence among the men and women so involved that Leoba's role as spiritual mother was not unique. One monk wrote a poem to commemorate "those, with which I designated your name, as that of my spiritual mother"[97]; others simply identify in passing references a woman as their "spiritual mother."[98] Two women, Eangyth and Heaburg, confided in Boniface their own understanding of the responsibilities of a spiritual director: "It is not so much thought on our own souls, but, what is more difficult and serious, thought on the souls of all those of different sex and age committed to us. We are to serve these many and varied minds and characteristics and afterwards to give an account before the judgment-seat of Christ not only for manifest sins in deed or in word, but also for those hidden thoughts which escape all men, and are plain to God alone." The women also reflected upon some of the commonsense reasons people had for seeking out a spiritual director: "Every person who is wanting in his own cause, and trusts not his own counsels, seeks out a faithful friend whose counsels he can trust, since he trusts not his own; he will have such faith in him as to reveal and lay open to him every secret of his heart." We see from these two comments that Eangyth and Heaburg perceived spiritual direction as it was perceived by the desert ascetics. It probed "those hidden thoughts which escape all," and it was conducted within a relationship of equality, trust, and faith.[99]

The second corollary of the desert ascetics' exploration of the human condition was the effect it had on Christian society's understanding of sin and how this in turn ironically affected spiritual direction. The starkness of the desert revealed to the ascetics the presence in each person's psyche of two warring spirits of good and evil, "one from God and the other from the devil."[100] If we want good to win "we must arm ourselves in every way against the demons. For they attack us from outside, and they stir us up from within," warned amma Syncletica, "so we must watch for the attacks of men that come from outside us, and also repel the interior onslaughts of our thoughts."[101] The attacks and onslaughts take the form of temptations, which, like passions, are neutral in themselves.[102] What we do with them is key, not whether we have them. Through temptation we are able to exercise the ultimate choice, the choice between good and evil. "It is only through many trials and temptations," amma Theodora explained, "that we can obtain an inheritance in the kingdom of heaven,"[103] for part of the Christian understanding of life is the priority it grants free will. Because "many are the wiles of the devil," there is need for spiritual direction in these choices to help us "direct our souls with discernment," for how else "are we to distinguish between the divine . . . and the demonic?" asked amma Syncletica.[104] Through careful examination of inner thoughts and with the help of someone one trust and has faith in, temptations can be openly discussed and choices made consistent with the goal of happiness.

This intense analysis of the inner life resulted in all ascetics realizing they were sinners. This realization was a step forward, however, for it kept them focused on the reward to be gained by choosing the good. "In the world, if we commit an offence, even an involuntary one, we are thrown into prison," amma Syncletica offered as an example; "let us likewise cast ourselves into prison because of our sins, so that voluntary remembrance may anticipate the punishment that is to come."[105] The deepened consciousness of sin also resulted in a profound sense of equality and community, and a lack of judgment within that community, and the realization that no one had the right condemn another. The desert spiritual director helped the novice come to this conclusion.

When monasticism left the desert and migrated to the West, this understanding of sin was modified gradually when it came into contact with the Germanic peoples and with developments within the institutional church. In the first centuries sin was primarily considered a breach in the relationship between Christians themselves and between the Christian community and God; repentance was the effort to reform oneself back to the image of God before the breach. Increasingly by the fourth century sin was thought of in more legalistic terms of transgression against a law, and repentance was perceived in terms of penalty, often performed publicly. These ideas were often codified into law and adopted by the church as part of the official penitential system. The system, though, proved too harsh, and many people turned instead to unofficial means to deal with sin.[106] They found holy people known for their spiritual discernment and revealed to them the state of their souls. We saw this occurring in the lives of Radegund and Hilda of Whitby. While in theory such spiritual direction was quite distinct from sacramental confession, it was not so in practice. Both were means to help achieve an end, salvation, which Christians believed was the whole purpose of earthly life. The association of spiritual direction and confession grew even stronger during the early Middle Ages once the Irish penitential system had borrowed much from early Eastern monasticism, particularly as interpreted by John Cassian.[107] Irish penance was performed within an one-to-one relationship and required a detailed recitation of sins and their nature, two details common to monastic spiritual direction. As the Irish practice was adopted by other Western peoples, two further modifications were made. The continental practice of private confession substituted the priest for the monk as the confessor. Since the monk is a layperson, and a priest is a member of an exclusively male clerical class, total linkage of spiritual direction with confession would have meant that women could not be spiritual directors any longer. Also, in an attempt to mete out remedies for transgressions fairly and equitably, the penitential book came into existence. Eventually this development had the unfortunate effect of once again turning the penitential system toward a legalistic rendering of penance. With such a rendering, theoretically the need for individualized spiritual direction would have diminished and the opportunity for direction decreased.

During the ensuing centuries, however, the opposite occurred, for two reasons. First, at the same time that confessional manuals were universalizing the human condition and categorizing behavior into rigid compartments, other aspects of society were individualizing human life and dissecting behavior according to intention.[108] People like Abelard and Heloise drew close connections among intention, sin, and the understanding of sin. "Wholly guilty though I am" of sexual misbehavior reasoned Heloise, "I am also, as you know, wholly innocent" because her intention was pure.[109] Such an argument defied attempts to categorized behavior without examination of interior motives. As a society, however, the medieval Christian West did not possess all the necessary tools for self-examination and needed guidance from those well-versed in self-examination; it needed spiritual guides. This need was compounded when annual sacramental confession was made mandatory in 1215. By the thirteenth century the penitential manuals had made intentionality part of the confessional discourse, thus again creating a need for guides capable of directing people in how to do so. As historian Jerry Root observes, "to make confession work" once it included an evaluation of intention, the church had to generate "a plethora of prescriptive, didactic writings to teach medieval penitents how to present themselves in confession."[110] A new space was being created in religious literature, and spiritual direction literature was among the numerous new genre that helped fill it. As in ages past women spiritual directors were numerous and accepted without question. Because spiritual direction now becomes literary as well as auditory, in the high Middle Ages we have direct access to women's spirituality and direction in their own words.

Chapter 4

The Great Medieval Directors

Some of the most remarkable examples of spiritual direction to survive from the high Middle Ages come from Hadewijch. Few facts are known about her life beyond the facts that she was an educated beguine who lived in the Low Countries during the thirteenth century. Contemporary sources take no notice of her, indicating her activity as spiritual director was not extraordinary or unusual. To the contrary, the historical context of Hadewijch's own letters reveals her directees and those reading her work after her death considered a woman in such a position as unexceptional. As a spiritual director she should be considered to be representative of the many women who also practiced spiritual direction within their communities without gaining notoriety. Her spirituality, though, is anything but representative. It is in context, in expression, and in influence unique. Few authors writing in any genre have been able to communicate so creatively and convincingly their own understanding of life as Hadewijch, and modern scholarship has barely begun to do her justice. Recent attention to her has already yielded major revisions in the history of mysticism and Hadewijch's contributions to it. When we insert the same basic facts in the history of spiritual direction, we must likewise revise the history of spiritual direction to place Hadewijch in a more prominent position.

Hadewijch left four different types of sources: letters, visions, poems in stanzas, and poems in couplets. No matter how different the form of her communication, though, all were written for the spiritual direction of younger beguines. There is little in the sources to indicate Hadewijch held an institutionalized position such as religious founder or monastic superior that would encompass spiritual direction, so we can assume it was her spirituality that drew others to her. "I did not speak thus because I wished to pray for you or to win your good will, but because you requested it,"

Hadewijch explained in a poem. "I answer you with pleasure, in simple terms, not a great length, concerning what you told me to treat of."[1] Requests were not the sole reason for her spiritual direction; she also believed it was her duty to help others learn what she knew. Now that "you have tasted me and received me outwardly and inwardly, and you have understood that the ways of union wholly begin in me," she is told in a vision by the divine, "lead all the unled."[2] She employed various methods in her spiritual direction, ranging from astute psychological analysis of human motivation to guidance by example. When instructing her directees about self-abandonment to Love, for example, she wrote plainly and to the point: "I wish to write something by which we may learn to recognize great marks of spiritual love, and also find a great example in what union she gave herself to Love. This was Mary Magdalen, who was one in unity with Love."[3] Even with both external urgings and an interior sense of vocation, Hadewijch never lost perspective of her role in direction. Primary guidance was always from God, and Hadewijch's was secondary: "I pray God that he may direct your understanding in his veritable Love, and that he may enlighten you with himself, and lead you by his deeper truth. For from me, you shall much lack this, although I also wished to speak for your profit."[4] She was always aware of the good such guidance could accomplish, knowing that "there were persons I delivered from sins, persons I delivered from despair."[5] She was simultaneously aware that any guidance she could give was limited by the very nature of the human condition: "Nobody who has loved Love with love could explain to others, or write, or bring to their understanding, all the wonders [she] finds in Love's sublimity."[6] The very fact that she recognized the limits of human communication and the need for direct personal experience, however, motivated her to try even harder to improve her direction, for "I could scarcely endure that anyone should love him less than I."[7] Her spiritual guidance started with this commonsense directive for attaining happiness:

> If you wish to experience this perfection, you must first of all learn to know yourselves: in all your conduct, in your attraction or aversion, in your behavior, in love, in hate, in fidelity, in mistrust, and in all things that befall you. You must examine yourselves as to how you can endure everything disagreeable that happens to you, and how you can bear the loss of what gives your pleasure; for to be robbed of what it gladly receives is indeed the greatest sorrow a young heart can bear. And in everything pleasant that happens to you, examine yourselves as to how you make use of it, and how wise and how moderate you are with regard to it. In all that befalls you, preserve your equanimity in repose or in pain. Continually contemplate with wisdom our Lord's works; from them you will learn perfection. It is truly fitting that everyone contemplate God's grace and goodness with wisdom and prudence: for God

has given us our beautiful faculty of reason, which instructs man in all his ways and enlightens him in all works. If man would follow reason, he would never be deceived.[8]

Echoing the desert ascetics Hadewijch placed tremendous responsibility on the individual in the search for perfection (which throughout her work she identifies as Love), even at the highest levels of mystical union. "For each revelation," Hadewijch argued in reference to her own development, "I had seen partly according to what I was myself, and partly according to my having been chosen." If she had not contributed accordingly she never would have arrived at the truth. This was confirmed in a vision where "the Voice said to me: 'O strongest of all warriors! You have conquered everything and opened the closed totality" though hard-won knowledge. "'It is right, therefore, that you'—whom the Voice calls the greatest heroine—'should know me perfectly.'"[9] It is through the strenuous exercise of reason that knowledge comes, and Hadewijch is not shy about emphasizing the role of reason in the pursuit of happiness. Her evaluation of reason is key to comprehending both her spirituality and her direction. Given that she is writing during the intellectual awakening of the West and its corollary establishment of the incomparable university system, her remarks are quite significant and revealing.

Reason, according to Hadewijch, is indispensable, for "he who wishes all things to be subject to him must himself be subject to his reason, above whatever he wills or whatever anyone else wills of him. For no one can become perfect in Love unless he is subject to his reason."[10] She admits "it is truly no easy risk to ask Reason's counsel about Love," yet it must be done, because "on this it depends to receive Love in her entirety."[11] Reason was needed to temper the alluring attraction of Love. "Love came to hold out to me the promise of all love," telling Hadewijch to pause and "'reflect that you are still a human being,'"[12] and reminding her that only "in winning the favor of Reason lies for us the whole perfection of Love."[13] The will is as dependent on reason as love, because "desire cannot keep silence, and Reason counsels her clearly, for she enlightens her with her will and holds before her the performance of the noblest deed."[14] Pleasure also needs the counseling of reason, because if left unattended, "Pleasure would certainly close her eyes and gladly enjoy what she possesses if fierce Desire, who always lives in fury, would tolerate it." Desire thus "awakens Reason, who says to Pleasure: 'Behold, you must first reach maturity!'" and refuses to be satisfied by incomplete Pleasure.[15] In a vision Queen Reason appeared to Hadewijch, along with three maidens. "'Do you know who I am?'" Queen Reason inquired. "'You are my soul's faculty of Reason, and these are the officials of my own household with whom you walk abroad,'" Hadewijch answered. The first maiden was Holy Fear, "who has examined my perfection"; the second maiden was "Discernment between you and Love, and she has tried to distinguish Love's will, kingdom, and good pleasure for yours"; and the third maiden was Wisdom "through whom I have acknowledged your power and your works when you let yourself be led by Love."[16]

After Queen Reason finished her lesson "she ordered me to acknowledge the whole number of my company; and I truly acknowledged it." Once Hadewijch realized reason had to be accompanied with holy fear, discernment and wisdom, "Reason became subject to me, and I left her." Then, and only then, "Love came and embraced me."[17] Hadewijch never inflated the position of reason. It was indispensable as a means to an end, not as an end in itself. Even as a means reason is fallible and must be considered warily. Sometimes reason errs, and "then when reason is obscured, the will grows weak and powerless and feels an aversion to effort, because reason does not enlighten it." For example, "reason well knows that God must be feared, and that God is great and man is small. But if reason fears God's greatness because of its littleness, and fails to stand up to his greatness, and begins to doubt that it can ever become God's dearest child, and thinks that such a great Being is out of its reach," then "Reason errs in this."[18] Hadewijch's "motive for telling you that reason errs"[19] was simple. It was to remind her directees not to rely solely in reason, for it was limited. Only when reason merged fully with love and the directee consequently did "lose yourself wholly in him with all your soul," only then "we shall reach our full growth."[20]

That, after all, is the purpose of life, to reach one's full growth. For Hadewijch God is actuality, and a human's potential is actualized only when fully united with God, "for he is in the height of his fruition, and we are in the abyss of your privation: I mean, you and I, who have not yet become what we are, and have not grasped what we have, and still remain so far from what is ours. We must, without sparing, lose all for all; and learn uniquely and intrepidly the perfect life of Love."[21] The person "who holds back anything in his heart cannot attain to the full growth of love."[22] There are many ways to interpret what Hadewijch means by *minne*, Love—God, Divine Love, the relationship of the soul to God[23]—but whatever connotation one wants to give the term, union with Love still remains her ultimate goal. Humans cannot rationally or emotionally deny this reality, "for interiorly Love draws them so strongly to her, and they feel Love so vast and so incomprehensible" that they can only experience "pleasure in proportion as Love was advanced or grew in themselves and in others" and "pain in proportion as Love was hindered or harmed."[24] Hadewijch urged her directees to "be satisfied with nothing less than Love. Give reason its time, and always observe where you heed it too little and where enough,"[25] but never forget that only "if you abandon yourself to Love, you will soon attain full growth."[26] Love alone "rewards to the full."[27] That means "one would choose or wish nothing except to desire above all what Love wills,"[28] because "all your perfection depends on this: shunning every alien enjoyment, which is something less than God himself; and shunning every alien suffering, which is not exclusively for his sake."[29] Only God can satisfy one's yearnings, because "in him is, in highest measure, eternal glory and perfect enjoyment."[30]

This human dependence on God is not arbitrary, Hadewijch is told in a vision, but integral to humanity's "noble nature, which makes you desire me in my totality."[31] At the center of this relationship is the freedom of all entities involved. Just as "Love knows no distinctions; [she] is free in every way,"[32] so, too, are humans. In fact, free will is absolutely essential to humanity: "I am a free human creature, . . . and I can desire freely with my will, and I can will as highly as I wish."[33] Humans, however, must understand "that neither in heaven nor in the spirit can one enjoy one's own will, except in accordance with the will of Love."[34] The way people learn this fundamental principle is the way Hadewijch learned it, through Jesus' "example and in union with him, as he was for me when he lived for me as Man."[35]

For Hadewijch the Incarnation holds the answer to life's most frustrating question, why suffering exists. The goal of life, as we have noted, is to be completed in the totality of the Divinity. What humans have to learn is how to "love the Humanity [of Jesus] in order to come to the Divinity, and rightly know it in one single Nature,"[36] because in the Incarnation "the Godhead has engulfed human nature wholly in itself."[37] Therefore, in order for one "to understand and taste him to the full," Hadewijch must suffer as he suffered.[38] Hadewijch herself prayed for this: "I desire that his Humanity should to the fullest extent be one in fruition with my humanity, and that mine then should hold its stand and be strong enough to enter into perfection until I content him, who is perfection itself." Because her ultimate wish is that "he might content me interiorly with his Godhead," she prayed for one gift in particular, "that I should give satisfaction in all great sufferings. For that is the most perfect satisfaction: to grow up in order to be God with God. For this demands suffering, pain, misery, and living in great new grief of soul,"[39] embracing her full humanity as Jesus did.

Hadewijch instructed her directees accordingly. "Even if you do the best you can in all things, your human nature must often fall short," so follow "unconditionally our Lord's guiding" and example.[40] Her directees "must be continually aware that noble service and suffering in exile are proper to man's condition; such was the share of Jesus Christ when he lived on earth as Man."[41] So, too, "with the Humanity of God you must live here on earth, in the labors and sorrow of exile."[42] Suffering should thus be freely chosen. "If you wish to be like me in my Humanity, as you desire to possess me wholly in my Divinity and Humanity," Hadewijch was told in a vision, "you shall desire to be poor, miserable, and despised by all men," because by such suffering "it will become so alien to you to live among persons" that she will be drawn away from incomplete happiness and instead "desire me in my totality."[43]

Rather than being meaningless, suffering to Hadewijch was the key to happiness. To the people seeking perfection "all pain is pleasant,"[44] even though human nature is such that it thinks the opposite is true. "For anyone who at the moment has no consolation presumes his life is leading nowhere; and anyone who has repose according to his pleasure presumes that he fully matches perfection. Thus the crowd are now deceived, and fancy they are perfect: that is untrue." On the other hand, "if they considered with reason," Hadewijch wrote, "and gave themselves to Love in suffering" instead, they would realize that suffering is a way of approaching and learning Love's ways. Love "will repay all pain with love," thereby transforming suffering into something positive. Hadewijch urged her directees to "come, desire to suffer in order to ascend so that we together in one knowledge may have fruition of our Love" and be led "into the blessedness that has been prepared in which Love shall be eternally."[45] Once someone realizes the purpose of suffering, to bring one fully into the humanity and love of God, then it ceases to be suffering, for "if he realizes in his mind that he suffers because of sublime Love, he will suffer gladly in every season." Likewise, "in every season suffering must befall him who wishes to serve sublime Love," if one desires "to experience that nature in which he is loved with love by Love."[46] Consequently, "the valiant heart that wills to suffer pain for Love, has no need of sadness," because suffering is actually the means by which the soul "shall know and understand all."[47]

Most important, to suffer is to imitate Christ in all aspects. "We all indeed wish to be God with God," Hadewijch astutely observed, "but God knows there are few of us who want to live as men with his Humanity, or want to carry his cross with him, or want to hang on the cross with him, and pay humanity's debt to the full."[48] Every time we rebel against suffering "we show plainly that we do not live with Christ as he lived."[49] Furthermore, by running away from suffering we risk missing his help, so Hadewijch encouraged her directees to "suffer gladly, in all its extent, the pain God sends you; thus you will hear his mysterious counsel."[50] Unless we die with Christ we cannot "rise again with him," which "to this end he must always help us."[51] Remember, Hadewijch urged, that "in each and every circumstance he was ready to perfect what was wanting on our part. And thus he uplifted us and drew us up by his divine power and his human justice to our first dignity and to our liberty."[52]

Suffering must not distract us or become an end itself, though—"do not be remiss in virtue, no matter what the suffering!"[53]—and with that in mind

Hadewijch offered much practical advice. She criticized one directee because "you busy yourself unduly with many things, and so many of them are not suited to you. You waste too much time with your energy, throwing yourself headlong into the things that cross your path." To overcome this Hadewijch entreated her "to observe moderation" and, more specifically, to obediently "follow the counsel I have given you."[54] To another directee just beginning the spiritual quest Hadewijch advised patience, since "your human nature must often fall short."[55] Because "you are still young and you must grow a good deal," Hadewijch had to remind her of the role of suffering: "It is much better for you, if you wish to walk the way of Love that you seek difficulty and that you suffer for the honor of Love, rather than wish to feel love."[56] Hadewijch had a balanced sense of a spiritual director's transitory role, however, and always urged her directees to remember that God is the true spiritual director. "In future do not desire the support of any person on earth or in heaven, no matter how powerful he is. It is as I told you, you are sustained by God, and you must wish to be supported by him, with great strength," Hadewijch wrote the same directee. "You must not doubt this, and you must not believe in men on earth, saints or angels, even if they work wonders." As long as Hadewijch is needed "I will tell you the help that is fitting for you," but she insisted that human spiritual direction was limited, however numerous the approaches may be.[57] "I entreat you, as a friend his dear friend; and I exhort you, as a sister her dear sister; and I charge you, as a mother her dear child; and I command you in the name of your Lover as the bridegroom commands his dear bride,"[58] Hadewijch reflected, but she knew that only by "following unconditionally our Lord's guiding" will "you be not wanting in the great works to which he has called you."[59]

As this detailed exposé clearly reveals, spiritual direction had matured by the high Middle Ages. In Hadewijch's hands it is self-conscious, articulate, and focused. There is much in Hadewijch's direction that is extraordinary. Her poetic and prose expression is quite impressive and accessible to the modern reader. While her lack of reliance on pietistic imagery or technical terminology may account for some of this accessibility, I would argue that the recent interest in Hadewijch is for a much different reason. She still has much to say to us today. The questions she asked and the answers she gave are timeless, and their relevance is universal. She spoke directly to the human condition, to the limitations and incompleteness of human nature, to the frustrations over life's superficiality, and to the restless search for totality, or, as she so often called it, the "hunger for Love."[60] She pondered the imponderables: suffering, the role of reason, the need for love, free will, and what it means to "remain a human being."[61]

Such a sophisticated spirituality was a highly significant contribution to Western culture, as contemporary scholarly activity has well proven.

Hadewijch was very much a part of what church historian Bernard McGinn identifies as "an important shift taking place in the thirteenth century: the earliest large-scale emergence of women's voices in the history of Christian thought."[62] While Hadewijch did not live to see her spirituality bear fruit, there is no doubt among contemporary historians of spirituality that it and the beguine world of mysticism within which her spirituality nourished had a tremendous impact on the shape of late medieval Christianity. What is even more significant to us here is that she communicated her spirituality through spiritual direction, affording us an almost unparalleled opportunity to appreciate both in a single form. We must remember, however, that as exceptional as the content of her spirituality may be, her direction simultaneously appears to be unexceptional, meaning that there is nothing in her sources or in any other primary sources to indicate that anyone considered her role as spiritual director to be unusual. The most unusual aspect about it seems to be that her direction was written and that the texts were preserved. This phenomenon allowed her spiritual direction to continue to guide many well beyond the original circle of directees,[63] even as her thought guides many in search of spiritual direction today.

While no other medieval women spiritual directors record their direction as thoroughly and directly as Hadewijch, there are innumerable examples of women spiritual directors. All sources reinforce the thesis that spiritual direction was an acceptable, unexceptional ministry for women. The sources also reinforce the thesis that women's spiritual direction of men was not extraordinary. Perhaps these theses are easiest seen proven in the direction of Hildegard of Bingen.

Of the some three hundred letters preserved from Hildegard's correspondence, the vast majority of them request, contain, or make reference to her spiritual direction. Strangers and close acquaintances alike wrote her for guidance on the basis of her renowned spiritual reputation, without concern that she was a "poor little woman," as Hildegard called herself.[64] Conrad, Archbishop of Mainz, longed "to hear your words of exhortation," because "we know that you have been inspired by the Holy Spirit."[65] Hermann, Bishop of Constance, introduced himself by telling her "the fame of your wisdom has spread far and wide and has been reported to me by a number of truthful people, reports that have made me desire to seek out your solace and support even from these far distant regions."[66] Christian, Archbishop of Mainz, confessed that "these obvious signs of your holy life and such amazing testimonies to the truth oblige us to obey your commands and to pay especial heed to your entreaties."[67] The monk of Amorach believed that "because you have ascended close to that brightness," Hildegard's guidance could "enlighten our minds so that our darkness may be dispelled by the rays of admonition, of exhortation, of chastisement."[68] After listing Hildegard's many accomplishments as a spiritual leader, Odo of Soissons confided that "this does not surprise us at all, because it does not exceed your purity and saintliness, without which no one can attain to such things."[69] The monk Geoffrey tells us that "I have heard of your reputation, which is spreading abroad.[70] One congregation of monks addressed her as "Hildegard, whom the Lord chose as His handmaid and made partaker of His multitudinous secrets" and prayed that "the angel of counsel" keep her unharmed.[71] One prior confessed of having "often heard of your fragrant and delightful reputation" because "you are so worthy, so pleasing, so lovable, so venerable to all those in whom He dwells that there is no question that He dwells in you."[72] All of her petitioners sincerely believed the counsel they could receive from her was invaluable because, as one congregation of monks stated it, "the truth of the Lord speaks through you." Convinced that Hildegard's insight into human nature was so keen that she could identify any elements capable of inhibiting a person from attaining perfection, these monks pleaded for her direction: "We humbly pray you not to hide anything from us which we should amend in ourselves. And since it has pleased the Lord to reveal many secrets to you, please hasten to reveal to us that which we ought to correct."[73]

While all these sources document clerical demand for Hildegard's spiritual direction, she did not limit her guidance to this one class nor solely to matters stemming from interior problems. Daniel, Bishop of Prague, decided that because "we have heard that your love has aided many people in their distress, we have been encouraged to seek the assistance of your prayers and advice, beloved lady, for we are shaken by secular tribulations."[74] The abbess Sophia claimed that the "Lord has inspired me, I believe, to lay down the heavy burden of administration which I bear, and to seek the seclusion of some little cell"; she desired Hildegard's guidance in this matter.[75] Some petitioners desired her direction in theological concerns. Eberhard, Bishop of Bamberg, wished "to submit the following matter" concerning Christology to be expounded "through your spiritual love,"[76] while Odo of Soissons wanted Hildegard "to resolve a certain problem for us" concerning God's paternity and divinity.[77] The monks of Amorbach were so concerned about the general state of religion among the people that they turned to Hildegard for help: "Now, O venerable mother, since you have frequently directed you admonitions to the clergy, as we know from your own writings, we ask as sincerely as we can that you direct them also to the laity." They believed such guidance would help since the laity "know that you speak on the authority of divine visions and commands," and, therefore, "they will incline their hearts more attentively to your words than they have to ours." Indeed, the condition of society was so bleak that the monks thought Hildegard's spiritual guidance was their only chance: "And so even though almost the whole world is overwhelmed by the darkness of error, that ray of ancient grace has shone in you, lest all the people perish."[78] Amalricus, Bishop of Jerusalem, confessed that he heard that "all your sisters" were being guided "under your supervision in Christ"[79]; he revealed this in a letter in which he requested spiritual direction for himself. No matter who the particular correspondence or what the specific occasion for the letter, in tone and approach, all the letters requesting direction have much in common with the sentiments expressed in the cleric Arnold's humble petition.

> Dearest lady, you know, not only from our conversations but also through divine revelation, how sorely my soul is tried by the tribulations of temptation, both inwardly and outwardly. Now, therefore, because I am weak and because I see in myself no progress at all toward God. I have written to you, my lady, fully prepared to follow your commands implicitly, whether founded in divine revelation or simply in the wisdom of your counsel. Do not spare me, I beseech you. Do not hide my iniquity from me. I fear for the tattered remains of my impoverished spirit, and I am terrified that I am being sucked down into the abyss.[80]

It is obvious from her correspondence that Hildegard had some very definite ideas about spiritual direction. First of all, she strongly condemned anyone who failed to guide people spiritually in accordance with the position he or she held. In a public sermon preached at Cologne and in a copy of the sermon sent in letter form to the Cologne clergy upon request, Hildegard tells us that "I have worn myself out for two whole years so that I might bring this message in person to the magistrates, teachers, and other wise men."[81] Second, she believed firmly that her direction was from God, "the director of all things."[82] Whereas Hadewijch used her visions as a vehicle to illustrate the correct direction, Hildegard held that her very "words do not come from a human being but from the Living Light."[83] In a letter to Arnold, Archbishop of Cologne, she insisted on the divine origin of her direction even more strongly. "Indeed, this very letter which I am now writing to you," Hildegard vehemently repeated, "contains nothing originating from human wisdom nor from my own will, but rather it contains those truths which the unfailing Light wished to reveal through his own words."[84]

There is also another difference between Hadewijch and Hildegard to consider, the position of the corresponding recipients of the direction. Hadewijch wrote to young beguines; Hildegard corresponded with popes, cardinals, bishops, and congregations. When writing to Pope Eugenius, for example, Hildegard softened the audacity of her chastisement with this qualifier: "This poor little woman trembles because she speaks with the sound of words to so great a magistrate. But, gentle father, the Ancient Man and Magnificent Warrior says these things. Therefore, listen."[85] Likewise, Hildegard added this explanation in a treatise she wrote after Eberhard, Bishop of Bamberg, requested her guidance in a theological matter: "Father, I, a poor little women, am able to expound upon the question you asked me, because I have looked to the True Light, and I am sending along to you the answer I saw and heard in a true vision—not my words, I remind you, but those of the True Light."[86] Hildegard's insistence on direct inspiration can also be attributed to the nature of her visions. Whereas Hadewijch's visions were ones in which she was "wholly melted away in him and nothing any longer remained of me of myself; and I was changed and taken up in spirit,"[87] Hildegard's were more didactic than ecstatic. When she offered spiritual direction she was just obediently writing about "those things I saw and heard in a true vision, while I myself was fully alert in mind and body."[88] Third, Hildegard believed that spiritual direction was essential to all who seek happiness. "Although you are on the way that leads to God," she explained to one abbot, "you are anxious in your spirit" and are "totally caught up in your own weariness." At such a time Hildegard advised him to "look to me" for guidance, and "I will help you,"[89] just as she herself is helped. Spiritual direction is also necessary because of the power of self-deception. "Your secret thoughts deceive you sometimes, and sometimes you are led astray by the taste of your own works," Hildegard warned the monk Godfrey, but if "the face of your desire turns toward me in joyful hope of recovery," then he will attain his goal.[90] It was Hildegard's worry about her own power of self-deception that led her to write Bernard of Clairvaux for guidance concerning "whether I should speak these things openly," that is, talk about visions which were "teaching me the profundities of meaning."[91]

While Hildegard's correspondence provides us with an opportunity to see the context, reasons, and judgments of her spiritual direction, her *Book of the Rewards of Life* is a summary of the knowledge she gained as a spiritual director. Historian Barbara Newman classifies it as a creative "synthesis of at least three medieval genres: the *psychomachia* or virtue-vice debate, the penitential and the other world vision," a synthesis that represents a compromise between an older static concept of human behavior and a newer psychological dynamism.[92] Her spiritual direction was in this sense unique, as was the book summarizing it. Hildegard's extensive use of visions and, in particular, her employment of highly visionary images, allegory, and rhetoric in her writings create a stumbling block for many modern readers.[93] Consequently, they render her spirituality and direction much less accessible to our nonvisionary mentality and make her *Book of Rewards* "rather tedious reading."[94] When patiently examined, though, we find Hildegard addressed the same type of ultimate concerns about the human conditions that Hadewijch did.[95] "Alas that I was ever created! Alas that I am alive! Who will help me? Who will free me?" lamented "a fifth image that had the form of a woman" and personified the vice of Sorrow of Time. "I listen to a lot of things from philosophers who teach that there is much good in God, but God does not do any good for me. If he is my God, why does he hide all his grace from me?"[96] Another voice from heaven responds with some pointed advice about pessimistic outlooks. "When the day rushes up to you, you call it the night; when salvation is present to you, you say that it is a curse; and when good things come to you, you say that they are evil." Rather, one should be as optimistic as the voice that Hildegard calls Heavenly Joy. "I do not do this. I give all my works to God because in some kinds of sorrow, there is gladness, and in some joy, prosperity. This is not like the day and night where one cannot be in the other."[97] Hildegard later reports the punishment for "those who had drawn the sorrow of time to themselves while they had been alive"[98]: purgation took place in "an arid and dry place that was surrounded with darkness" where "they suffered the contrariness of this darkness."[99] The voice from heaven interrupts this vision with final advice on how to avoid deprivation of the Living Light: "If they want to escape its punishments, let them turn to the spiritual life while they are alive. Or if they are already on a spiritual journey, let them follow the ordinary common strictness, let them submit themselves to humble obedience, and let them contemplate those Scriptures that bring them heavenly joy. However, let them not do all these things boldly, but only under the direction of their spiritual advisor."[100]

Hildegard's presentation and method of spiritual direction throughout the book is simple yet highly structured. Thirty-five vices in all first appeared in images and make a statement, then a virtue responded to a vice's statement, then a vice explained in more detail, and, finally, a vision appeared describing the punishment due such a vice. When the vice made its case for the first time, Hildegard presented it in nontechnical language reminiscent of Everyman's problems with the human condition. Gluttony said, "God created all things. How then can I be spoiled by all these things? If God did not think these things were necessary, he would not have made them. Therefore I would be a fool if I did not want these things."[101] Impiety stated its case in terms of free will: "I do not want to obey either God or man. For if I were to obey either of them, he might order me to do something best for him and not for me."[102] To Strife, self-defense is at its heart: "As long as I breathe and live, I will not allow anyone to strike me with the madness of his will."[103] Despair justified itself thus: "I have been greatly frightened! Who can console me? Who can help me and rescue me from these calamities that oppress me?"[104] And so Hildegard continued, discussing every vice she was aware of and offering her insight as to how to find positive answers to the questions asked in negative voices. As with the desert ascetics and Hadewijch, Hildegard believed that people must resolve their interior frustrations in the face of the mystery of life before true happiness could be attained. Her spiritual direction was meant to help directees do that.

With Hadewijch and Hildegard we have two extremes as examples of women spiritual directors. Hadewijch was a relative unknown, apparently with no widespread reputation for sanctity, and exerted little to no influence in her own lifetime. Hildegard, renowned throughout the West for her prophetic insight and spiritual direction, was eagerly sought out by her contemporaries of all social classes and status. Hadewijch was a participant in an innovative, marginal lay movement, whose purpose was frequently opposed and often challenged, and her life probably ended in exile from that movement. Hildegard was a powerful abbess of the Benedictine order, the oldest and most popular major religious order in the West. That the near-invisible Hadewijch had a developed sense of vocation as a spiritual director indicates that a woman spiritual director was not an anomaly, nor was the position in anyway reserved for the extraordinarily gifted or renowned. That the highly visible Hildegard likewise committed herself to a ministry in spiritual direction eliminates the possibility that women spiritual directors could practice only on the fringes of society. These conclusions are reinforced by an overwhelming amount of evidence we have of women spiritual directors practicing and influencing openly and extensively with societal approval throughout the whole of the high Middle Ages. Our knowledge of spiritual direction is, unfortunately, limited to members of religious groups by numerous historical realities. First of all, we are relying solely on written sources, and the religious had a near monopoly on literacy until the late medieval period. Second, those seriously seeking an understanding of life's mysteries in spiritual terms would be most likely attracted to religious life. Consequently, religious communities were environments that encouraged spiritual directors and acknowledged the need for spiritual direction. Third, the understanding developed in early monasticism that abbots and abbesses functioned both as administrators and spiritual directors was still embraced in the high Middle Ages. Indeed, if an abbess was not guiding her nuns toward spiritual ends, contemporaries would consider that to be amiss.

This last point can be seen in the testimony of Peter the Venerable in a letter written to Heloise and in Abelard's letter to Heloise outlining his ideas for a monastic rule for women. Because an abbess must assume "the care of bodies as well as of souls,"[105] Abelard stated that she must always remember her primary "care is for spiritual rather than material matters."[106] Peter the Venerable discussed the abbess' role as spiritual director at greater length with Heloise, to whom, he confessed, he was "drawn to you by what many have told me about your religion."[107]

You are indeed the disciple of truth, but in your duty towards those entrusted to you, you are the teacher of humility. For surely the teaching of humility and of all instruction in heavenly matters is a task laid on you by God, and so you must have a care not only for yourself but for the flock in your keeping; and being responsible for all shall receive a higher reward than theirs. Yes, the palm is reserved for you on behalf of the whole community, for, as you must know, all those who, by following your lead, have overcome the world and the prince of the world, will prepare for you as many triumphs and glorious trophies before the eternal King and Judge. Moreover, it is not altogether exceptional amongst mortals for women to be in command of women, nor entirely unprecedented for them even to take up arms and accompany men to battle. . . . For you will make honey, but not only for yourself; since all the goodness you have gathered here and there in different ways, by your example, word, and every possible means, you will pour out for the sisters in your house and for all other women.[108]

Thomas of Celano's vita of Clare of Assisi presents us with a detailed description of how Clare fulfilled her role as spiritual director. "Now, as she was the teacher of the unlearned and, as it were, the directress of the maidens in the palace of the Great King, she molded them with the best training," Thomas began. First "she taught them to shut out all tumult of this earth from their minds"; next, "she would teach them not to be influenced by love of kinsfolk"; after that, "she exhorted them to contemn [*sic*] the demands of the body and to subject the conceits of the flesh to the control of reason"; finally, "she desired them to work with their hands at definite hours."[109] Thomas of Celano also reported that Clare "began to enlighten the whole world,"[110] in particular Hugolino Segni, "both when he was Bishop of Ostia and later when he had been raised to the Apostolic See." He frequently "would call upon her by letter for assistance, and always felt that she helped him."[111] We have a few examples of Clare's direction in her letters to Agnes of Prague and Ermentrude of Bruges. Five in all, each letter contains some spiritual direction, sparingly expressed because Clare does not want to "weary you with needless speech."[112] Her direction is filled with exhortations to imitate Christ. "O dearest one, look up to heaven, which calls us on, and take up the Cross and follow Christ, Who has gone on before us: for through Him we shall enter into His glory after many and diverse tribulations," Clare wrote to Ermentrude. "He will be your help and best comforter for He is our Redeemer and our eternal reward."[113] She used the mirror as a simile for *imitatio Christi* in her letters to Agnes, urging her to "continually study your face within it"[114] to be assured of becoming God-like. "Place your mind before the mirror of eternity!" Clare wrote: "and *transform* your whole being *into the image* of the Godhead Itself through contemplation! So that you too may feel what His friends feel as they taste *the hidden sweetness* which God has reserved from the beginning for those who love Him."[115]

The holy women of the renowned monastery at Helfta—Gertrude of Hackeborn, Mechthild of Magdeburg, Mechthild of Hackeborn, and Gertrude the Great—provide us with an example of how spiritual direction helped establish a women's monastery as a renowned center of spirituality for generations. With the election of Gertrude of Hackeborn as abbess in 1251, Helfta began its journey. Although Gertrude left us no sources, it is clear that her spirituality and direction established an environment conducive to the creation of a vibrant new spirituality. A large portion of that environment was due to her insistence that the women utilize all aspects of their created being, and to that end she insisted they all train "the high point" of their intellect in a school that encouraged secular knowledge as well as spiritual understanding.[116] After years as a beguine, Mechthild of Magdeburg, author of the influential *The Flowing Light of the Godhead*, went to the Helfta monastery in search of good spiritual direction; she hoped that God might direct her to "someone who would be able to help her to make progress in her spiritual life." She found her prayers answered in the person of Gertrude of Hackeborn, and in a "discourse, full of the authentic sweetness of the Holy Spirit" with the abbess's sister, Mechthild of Hackeborn, author of the highly popular *Book of Special Graces*.[117] Gertrude the Great often sought out Mechthild of Hackeborn's spiritual direction. Once "Gertrude besought Dame Mechthild to petition the Lord for her, especially for the virtues of gentleness and patience"; another time she "humbly begged" Mechthild "to ask the Lord about the gifts" Gertrude the Great had recently received. "Dame Mechthild, as she had been asked, took counsel with the Lord" and received this extraordinary promise: "And the Lord added: 'Each time she wants to speak with others, she should draw into her soul a deep breath of inspiration from my divine heart. Whatever she says will then be spoken with certainty; neither she nor those who hear her can be deceived, because the secret intentions of my divine heart will be revealed in her words."[118] Gertrude the Great used this gift to its fullest potential in her spiritual direction: "Many felt the effects of grace not only in her counsels but also through her prayers." In addition, "many have testified to having been more profoundly moved by a single word of hers than by a long sermon by any of the best preachers," quite a telling comparison.[119] Certainly she took her gift of spiritual direction seriously. She wrote and published *The Herald of Divine Love* because she saw it as a form of spiritual direction. "'When [the people] hear about these graces that you have received, others may be brought to desire them for themselves, and by thinking about them, they may try somewhat to amend their lives,'" she reported the Lord telling her.[120] "Just as students attain logic by way of the alphabet,"

Gertrude the Great explained more poetically in Book Two, "so, by means of these painted pictures, as it were, they may be led to taste within themselves that hidden manna."[121]

At the heart of all her "painted pictures" was the proper exercise of free will. Free will is the means all humanity possesses to achieve the end goal of happiness. "I have given to everyone a golden tube of such power," said the Lord in a vision to Gertrude, "that he may draw whatever he desires from the infinite depths of my divine heart." She understood the tube to mean free will, "through which a man may claim for his own every spiritual good, both heavenly and earthly." Gertrude was quick to claim her happiness, as she answered, "'See, Lord, here is my heart, empty of all creatures. I offer it to you with my whole will'" and gave thanks to God for having "a will of such nobility that with it [a person] can gain infinitely more than the whole world could achieve with all its powers."[122]

Angela of Foligno presents us with yet another kind of spiritual direction. Not a member of a traditional monastery or a marginal group, Angela began her spiritual quest as a "woman of lay state, who was bound to worldly obligations, a husband and sons, possessions and wealth, who was unlearned and frail"[123]; she ended her life as a Franciscan tertiary. She gained renown in her own day as a spiritual director, telling her male directees "and all the others who are not here" at her deathbed that "it pleased the divine goodness to place in my care, under my solicitude, all his sons and daughters who are in the world, who are on this side of the sea or beyond it."[124] The most famous spiritual director of the period was Ubertino of Casale, the controversial leader of the Spirituals in the internal Franciscan turmoil. He met her when he was twenty-five, and he recorded the effect her spiritual direction had on his life: "She restored a thousandfold all those spiritual gifts I had lost through my sins; so that from that moment I have not been the same man I was before. The splendor of her radiant virtue changed the whole tenor of my life. It drove out the weakness and languor from my soul and body and healed my mind torn with distractions."[125] For the most part, however, her directees remain unidentified, beyond the fact that among them were both men and women.

Angela's *Book* is composed of two parts. The *Memorial* is a lengthy description of the steps she took in her search for perfection, while *Instructions* concentrates more on her spiritual direction to others desirous of perfection. As such, *Instruction* includes letters, guidance, and Angela's personal reflections on life's mysteries. Her opinion of her spiritual direction is humble. She claimed her letters contained only "very ordinary words" that were "useless unless one has a true knowledge of God and self." The only possible benefit she could see from them was if "they bestow upon you the kind of knowledge I am referring to," and for this ability she prayed "to God to give us this light."[126] To the question of why there is suffering "she replied, 'One must know God and oneself.'" Without knowledge of self one cannot know God. "The more one knows, the more one loves; the more one loves, the more one desires; the more one desires, the more one grows in the capacity to act accordingly," Angela argued, "and the true test of pure, true, and upright love is whether one loves and acts in accordance with the love and action of the loved one." This is difficult, since the God-man "always loved, and always practiced poverty, suffering and contempt. Therefore, like Christ, the one who loves him should always love, put into practice, and possess these three things."[127] To another directee she explained further: "For certainly, my dear son, these three things are the basis and the fulfillment of all perfection. For in these three, the soul is truly enlightened, perfected, purged, and most fittingly prepared for divine transformation."[128]

Because it may at first be difficult to grasp the truth Angela provided a metaphor to explain the process leading to happiness in a letter to yet another directee. "Just as hot iron put into the fire takes on the form of the fire—its heat, color, power, force—and almost becomes the fire itself, as it gives itself completely and not partially while retaining its own substance, similarly the soul united to God and with God by the perfect fire of divine love, gives itself, as it were totally and throws itself into God. Transformed into God, without having lost its own substance, its entire life is changed and through this love it becomes almost totally divine. For this transformation to happen, it is necessary that knowledge comes first, and the love follows which transforms the lover into the Beloved."[129] As a final help to her directees in their search for happiness, she reminded them continually of the benefits of prayer. "No one can be saved without divine light" because it alone "leads us to the summit of perfection. Therefore if you want to begin and to receive this divine light, pray. If you have begun to make progress and want this light to be intensified within you, pray. And if you have reached the summit of perfection, and want to be super illuminated so as to remain in that state, pray."[130] Indeed, all the virtues necessary for happiness may be obtained by prayer: "If you want hope, pray. If you want charity, pray. If you want obedience, pray. If you want chastity, pray. If you want humility, pray. If you want meekness, pray. If you want fortitude, pray. If you want some virtue, pray." Angela then offered some insightful guidance about what one must do "when the soul wants to improve its prayer." The soul must "enter into [prayer] with a cleansed mind and body, and with pure and right intention," as well as "turn evil into good." After putting one's soul "to a kind of scrutiny," the person will "begin to feel the presence of God more fully than usual." Angela thus directed all her directees to an intense life of prayer, believing firmly that "it is through prayer, then, that one will be given the most powerful light to see God and self."[131] It is of no wonder that the writer of an epilogue attached to some manuscripts of the *Book* urged the readers to "learn from Angela the great counsel" and "teach it to men and women and all creatures" what Angela so diligently revealed: "Angela showed us not only that Jesus' way was possible and easy, but also that it leads to the highest delights."[132]

We must remember that discussion here of women spiritual directors has been representative, not exhaustive. I have chosen women from different social groups, from different regions, and with different degrees of fame to emphasize that as spiritual direction became more self-conscious, explicit, and specialized during the high Middle Ages, women continued to participate in the ministry at all levels of society. Many, many more names could be added to a list of medieval women spiritual directors, some famous throughout the West in their contemporary world and well respected, others known only to us through a passing reference in a source. Letters from Clare of Assisi to Agnes of Prague, for instance, were quoted as evidence of Clare's ministry in spiritual direction, but their correspondence also makes clear that Agnes was a spiritual director in her own right.[133] On the other hand, the vitae of saints like Marie d'Oignies. Margaret de Ypres, Beatrice of Nazareth, Elisabeth of Spaalbeek, Ida of Louvain, Ida of Nivelles, Juliana of Mont-Cornillon, and Ivette of Huy all refer to the women engaging in spiritual direction to some degree.[134]

There is one common characteristic, however, that these women shared. All their spiritual direction was influential. Although relatively unknown in her day outside of a small circle of religious who ultimately rejected her, Hadewijch exerted significant influence on future German mysticism. Specifically, scholars document the influence Hadewijch had on Meister Eckhart and John Ruusbroec, who, theologian Louis Bouyer claims, did "convey and effectively prolong her heritage, during centuries when she herself remained practically unknown."[135] Hildegard of Bingen's personal spiritual direction influenced many members of the hierarchy of her day, while her written direction has enjoyed continuous influence throughout the centuries. Angela of Foligno's works, influential among Franciscans throughout the ages, played a significant role in the development of early modern Spanish and French schools of theological thought. With the women of Helfta we see the rise of a new phenomenon in the West, the powerful woman mystic who willfully bypassed hierarchical direction and established their authority and their source of guidance directly in God. The Helfta women's individual works each have their own history of influence. In 1603, for instance, we have the personal testimony of Bishop Diego of Tarragona, confessor to Philip II of Spain, concerning Gertrude the Great's *Insinuationes Divinae Pietatis:* "I cannot tell you with what joy and consolation I have received the book of the Revelations of Gertrude. . . . This was at the time when Philip II was seized with his last illness. I considered the works so suitable for his circumstances, that I read it to him frequently; and he found such consolation and sweetness therein, that in my absence he caused it to be read to him by the Infanta Madame Elizabeth."[136]

More important, each of the women spiritual directors provided her immediate society with answers in the pressing inquiry into the human condition. Hadewijch urged all to appreciate reason, to love Love, and to aspire to the fullness of one's potential; she offered an explanation of suffering. Hildegard believed happiness came by eliminating vices, and to that end she dissected those vices and promoted communal support in the form of spiritual direction as a way to attain that happiness. Clare of Assisi claimed that humanity's transformation into the Eternal, into *imago Dei*, was the key to happiness, while free will was the essential factor for Gertrude the Great. Angela of Foligno believed the human condition enjoyed happiness when self-knowledge led one to knowledge of God. Some of the women wrote of these matters plainly, and thus their message is relatively accessible to us today. Some wrote in highly stylized, flowery, pietistic language, and their message is more difficult for us to grasp. Regardless of the written style, however, the spiritual direction was obviously effective for the intended recipients, their directees, and that really was the only goal of these women spiritual directors. The fact that they were women was not a hindrance or limitation to either directors or directees, or to society in general. Peter of Dacia's description of his quest for a spiritual director appears typically oblivious to gender issues. When he prayed to God to "show me some one of his servants from whom I might learn the ways of his saints," God responded by "show[ing] me many persons of both sexes."[137] Eventually he recognized Christine of Stommeln as the best director for him, never once intimating any hesitancy or surprise that his prayer was answered in the person of a women. Apparently the reminder of the author of Angela of Foligno's epilogue was unnecessary: "Remember, most dearly beloved ones, that the apostles, who first preached Christ's life of suffering, learned from a woman that his life was raised from the dead."[138]

Chapter 5

Late Middle Ages:
Direction Comes of Age

During the late medieval and the early modern periods in Western Christianity, spiritual direction came into full maturity. Women are at the center of the maturation process, particularly women mystics. For a few decades now we have seen scholars give women like Julian of Norwich, Birgitta of Sweden, Catherine of Siena, and Catherine of Bologna their long-overdue recognition as leaders of the crucial mystical movement that swept through Western society; we do not yet see them also identified as pivotal leaders in spiritual direction. This is unfortunate, for the latter helps explain the former. The mystical movement became such a pervasive phenomenon during the period because mystics were able to direct so many people toward it. Women mystics were not the only women spiritual directors, but since they were ubiquitous and left numerous sources we will start our discussion with two of the most well known to us today, Catherine of Siena and Julian of Norwich. With Catherine we are singularly fortunate to have both her thoughts about spiritual direction in theory and a record of her actual direction. The *Dialogue* contains discussion of the limits, goals, and methods proper to good spiritual direction as well as a thorough explanation of the spirituality she communicated; her letters supply us with specific examples of how she guided people in need of direction.

Catherine is remarkably conscious of her role as spiritual director, so her discussions about it are explicit and to the point. She began her exposé of the ideal director by asking God to elaborate on three issues that affect spiritual direction "so that I may be able to serve you and my neighbor" and "not fall into any false judgments about your creatures" when offering them guidance. "Sometimes people will come to me," Catherine reported, "asking for counsel in their desire to serve you and wanting me to instruct them." She was confused, though, about the proper direction to give when she found "this one spiritually well disposed" and another "to have a darksome spirit." Specifically she wanted to know "should I or can I judge the one to be in light and the other in darkness? Or if I should see one going the way of great penance and another not, should I judge that the one who does greater penance is more perfect than the other?" The second query pertained to discernment: How is a spiritual director "to tell whether a spiritual visitation " is true or false?[1] God's response to these issues began with a review of the fundamental Christian understanding of the human condition. First, Catherine must remember to "rise above" her senses "so that you may more surely know the truth" and be satisfied by God, remembering at all times that "no one can walk in the way of truth without the light of reason that you draw from me, the true Light, through the eye of your understanding. You must have as well the light of faith." Happiness follows logically. "If you exercise this faith by virtue with the light of reason, reason will in turn be enlightened by faith, and such faith will give you life and lead you in the way of truth. With this light you will reach me, the true Light; without it you would come to darkness." Concretely, what this means is that "you must all be enlightened to know the transitory things of this world, that they all pass away like the wind. But you cannot know this well unless you first know your own weakness." Self-knowledge includes recognition of the inclination to sin, although it is "not that this law can force any one of you to commit the least sin unless you want to," because all possess free will. The very existence of this inclination to evil, though, "is reason for you to learn to know yourself and to know how inconstant is the world." Like so many spiritual directors before her, Catherine believed that self-knowledge was a key to happiness and that "lack of knowledge is the cause of [people's] evil." Such ignorance prohibits recognition of "the dignity they should preserve in themselves."[2]

"Perfect souls," God reveals to Catherine, "must truly know themselves and me." It is precisely "because they return continually to the valley of self-knowledge" that "their exaltation and union with me is never blocked."[3] Perfect souls also must realize that it is "I, gentle first Truth" who names "the situation, the time, and the place, consolations or trials, whatever is necessary for salvation and to bring souls to the perfection for which I chose them."[4] Only the soul who "has come to know my will" and "attends only to how she may keep and intensify her perfection" will succeed. What is perfection? It "is when she came to know when she saw this gentle loving Word, my only-begotten Son." After such knowledge is attained the soul "has no eyes for herself, for seeking her own spiritual or material comfort. Rather, as one who has completely drowned her own will in this light and knowledge, she shuns no burden from whatever source it may come." The perfect soul's "chief desire ought to be to slay your selfish will so that it neither seeks nor wants anything but to follow my gentle Truth." Once the perfect so control their will "they are always peaceful and calm" and "find joy in everything."[5]

The reason why God began his answer to Catherine's queries concerning spiritual direction with such an extensive review is because this is where all spiritual direction must start. "I told you," God explained, "the teaching and principal foundation you should give to those who come to you for counsel because they want to leave behind the darkness of deadly sin and follow the path of virtue. I told you to give them as principle and foundation an affectionate love for virtue through knowledge of themselves and of my goodness for them."[6] Once the spiritual director communicates these general principles, then she may turn her attention to more specific issues. Concerning the matter of judging another, God offered "three specific things I want you to do so that ignorance will not stand in the way of the perfection." First, "your judgment should always be qualified." The good director "should correct their bad habits in a general way and lovingly and kindly plant the virtues."[7] Second, "you neither should nor can assume" that any directee "is guilty of serious sin, for often as not your judgment would be false." Sometimes a person appears to be in darkness, but it is due not to his or her will, but "because I, God eternal, have withdrawn." Thus, "how foolish and reprehensible would be any judgment you or anyone else might make by assuming on appearances alone, just because I had revealed to you the darkness that that soul was in, that is was because of sin." Last, rather than indulging in such speculation, good spiritual directors should "concentrate on coming to know yourselves perfectly, so that you may more perfectly know my goodness to you. Leave this and every other kind of judgment to me, and take up compassion."[8]

Still, when a good director sees "something that is clearly sinful," she may first try to "correct it between that person and yourself." If the person "refuses to change, you may reveal the matter to two or three others, and if this does not help, reveal it to the mystic body of the holy Church." This is a subtle reminder of one of Catherine's persistent themes, that all actions of the individual have an effect on the community,[9] for there is no act of virtue and "no sin that does not touch others."[10] This principle is fundamental for all Catherine's spiritual direction: "Keep in mind that each of you has your own vineyard. But every one is joined to your neighbors." Thus, as insistent as Catherine was that spiritual direction respect the individual, she was equally insistent that the community be taken into account: "All of you together make up one common vineyard, the whole Christian assembly, and you are all united in the vineyard of the mystic body of holy Church from which you draw your life."[11]

To Catherine's corollary question of whether the spiritual director should judge penance to be a sign of greater perfection, the answer is basic: "I do not want you, dearest daughter, to impose this rule [of penance] on everyone. For all bodies are not the same, nor do all have the same strong consitutions."[12] Good spiritual directors will individualize and give penance to directees "as an instrument but not as their chief concern—not equally to everyone but according to their capacity for it and what their situation will allow."[13] The second issue Catherine wanted addressed, that pertaining to discernment of true or false spiritual experiences, is likewise simple. The sign of the true experience is gladness in the soul, and "you can discern that gladness is indeed signaling a visitation from me: if it is joined with virtue."[14]

According to Catherine spiritual direction is the obligation of all Christians. Because "it is your duty to love your neighbors," it follows logically that "you ought to help them spiritually with prayer and counsel, and assist them spiritually and materially." All Christians "owe each other help in word and teaching and good example, indeed in every need of which you are aware, giving counsel as sincerely as you would to yourself."[15] This mandate is grounded in the mystic body of the church, a doctrine upon which clerical responsibility rests, as well as in spiritual equality. In the fourteenth century Catherine's statement that God is "not a respecter of persons or status but of holy desires"[16] was radical in its implications, as was her constant reminder that all people were free. The soul of every person "is free" and "cannot be dominated unless she consents to it with her will, which is bound up with free choice."[17] This "freedom of your humanity" is at times overwhelming in its possibilities, but absolutely essential to happiness, because "you were freed from slavery so that you might be in control of your own powers and reach the end you were created for."[18] That end, knowledge of the Word, allows one to "share in the eternal Godhead made one with humanity, whence you will draw that divine love which inebriates the soul."[19] It is the undebatable and inescapable destiny of all persons. "Know that no one can escape my hands, for I am who I am, whereas you have no being at all of yourselves. What being you have is my doing; I am the Creator of everything that has any share in being," God told Catherine.[20] "And more than this—can you see?—there this union of the divine nature with the human. God was made human and humanity was made God."[21]

Thus Catherine offered her directees traditional rather than original spirituality. Union with the Godhead brought perfection, and, therefore, happiness, the goal of humanity. For that end "I will make my Son a bridge by which you can all reach your goal," God told Catherine.[22] "You must all keep to this bridge" at all times, "following in the footsteps of this gentle loving Word" because "there is no other way you can come to me."[23] That is why "I had created [Adam and Eve] in my image and likeness, so that they might have eternal life, sharing in my being and enjoying my supreme eternal tenderness and goodness."[24] Unfortunately, "by Adam's sinful obedience"[25] the road was so broken up that no one could reach everlasting life," until "I gave you a bridge, my Son, so that you could cross the river, the stormy sea of this darksome life, without being drowned."[26] Suffering became redemptive with this bridge. Since humanity "was incapable of atoning for sin," God explained, "your nature had to be joined with the height of mine, the eternal Godhead, before it could make atonement for all of humanity. Then human nature could endure the suffering, and the divine nature, joined with that humanity, would accept my Son's sacrifice on your behalf," all so "you might in truth come to the same joy as the angels."[27] Indeed, "in this way and in no other is suffering of value."[28]

When we turn to Catherine's letters we see these principles applied to specific situations. In one letter to the abbess and the subprioress at the Sienese monastery of Santa Marta, for example, Catherine began her counseling by praying for "the grace of knowing both ourselves and God" because perfection cannot come "without dwelling within the cell of our heart and soul where we will gain the treasure that is life for us." Proper exercise of one's free will is addressed next. Catherine informed her directees that "I want you to want things to go not your own way, but the way of one who is. You will then be stripped of your own will and clothed in his," and thus "will accept this circumstance and every other, however difficult, realizing that it must be for our good." Obedience to those in authority, including God, will further ensure the proper exercise of free will. In the abbess and subprioress's case this meant accepting "the burden [God] has placed upon you, the burden of governing his little sheep." Any attempt to evade "hard work under the plea that these things are temporal" is wrong because "all things come from the highest Goodness." Since these women were the spiritual directors of the other women in their monastery, like Catherine they must "have concern for your sisters' souls." Catherine repeated her rules for spiritual direction. They must guide each person individually and if in error "see to it that they are punished according to each one's capacity. If a person is in a condition to carry ten pounds, don't load twenty on her, but accept from her as much as she is able to give." Above all else, "dearest sisters, don't let correction seem burdensome to you," for spiritual direction is an obligation "rooted in the holy virtue of charity."[29]

Other principles of Catherine's spirituality and spiritual direction are found in a letter to two members of the Order of Mantellate of Siena, Monna Giovanna di Capo and one Francesca. Reminding them of the "dignity which you received from God" when he gave "us, *us*, his own image and likeness—just so we might experience and enjoy him, and share in his eternal beauty," Catherine proceeded to help the women accept their trials and tribulations by reviewing the fundamental Christian understanding of free will and suffering. "He gave us our will to love that will of his. The will of the Word wants us to follow him on the way of the most holy cross by enduring every pain, abuse, insult, and reproach for Christ crucified, who is in us to strengthen us." Initially this is difficult, but "once we have seen and understood that God wants only our good," it becomes easy. We even "rejoice and are content with whatever God permits: sickness or poverty, insult or abuse, intolerable or unreasonable commands. We rejoice and are glad in everything, and we see that God permits these things for our profit and perfection." Life's mysteries and "knowledge of ourselves" were "hidden to our coarseness until the Word, God's only-begotten Son, became incarnate, but once he had chosen to be our brother, clothing himself in the coarseness of our humanity, it was revealed to us." It is only through the bridge, God's son, that "love transforms and makes the beloved one with the lover."[30]

Catherine's directees formed a wide-ranging group, from royalty to relations to lay people, but with each her guidance was consistent with her spirituality and principles for spiritual direction. She was intent upon making these accepted by the whole church, and to that end she even lectured the pope about his failure to fulfill his obligation as spiritual director. "Those who are in authority," Catherine began in a letter to Pope Gregory XI with a thinly veiled reference to the pope himself, "do evil when holy justice dies in them because of their selfish self-centeredness and their fear of incurring the displeasure of others. They see those under them sinning, but it seems they pretend not to see and do not correct them." This failure to guide "is why those in their care are all rotten, full of uncleanness and evil!" And a person who acts thus "not only does he fail to rescue his little sheep from the clutches of the wolf; he devours them himself." Catherine then, with inescapable directness, addressed her remarks specifically to Pope Gregory: "I hope, by God's goodness, my venerable father, that you will snuff this out in yourself!" Her demands became even stronger. "I want you to be the sort of true and good shepherd who, had you a hundred thousand lives, would be ready to give them all for God's honor and other people's salvation!"[31] Catherine was not advocating indiscriminate condemnation of people by the hierarchy, only the proper condemnation of sin wherever it be found. "We mustn't love the vice we see in others, but we should love and reverence the person," Catherine directed Nicolo Soderini. "As for their sins, leave the judgment and chastisement to God."[32]

The themes of self-knowledge, free will, *imitatio Christi,* individuality, and communal responsibility are present time and again in her direction. Communal responsibility is particularly important for those in authority. "If virtue is so necessary for every one of us, if each of us individually needs it for the salvation of our soul," Catherine reminded Pope Gregory, "how much more do you need this constancy and strength and patience—you who must feed and govern the mystic body of holy Church."[33] To Charles V, King of France, she demanded complete *imitatio Christi* for the sake of his realm. "So I am asking you and I want you to follow Christ crucified," she implored, "and be a lover of your neighbors' salvation. Show you are a follower of the Lamb, who in hunger for his Father's honor and the salvation of souls chose his own physical death."[34] Phrases echoing the sentiments expressed in the *Dialogue* are strewn throughout the letters. "I want you to gain knowledge of yourself, without confusion, from the darkness. And from your good will I want you to gain knowledge of God's infinite goodness," she instructed her own spiritual director, Raymond of Capua, after their relationship became reciprocal, and she became his spiritual director. "I don't want you to yield to weariness or confusion, no matter what may trouble your spirit. I want you to keep the good, holy, and true faithful will that I know God in his mercy has given you."[35] She reminded the Archbishop of Otranto, Iacopo da Itri, that "you know well that neither the devil nor anyone else can force our will to the least sin," just as she encouraged Piero Gambacorta to "open wide your eye of self-knowledge,"[36] as she did Pope Gregory XI, Monna Giovanna, Raymond of Capua, Bernabo (Visconti of Milan), and just about everyone she directed. Underlying all her guidance, though, was the fundamental premise of Christianity: Happiness is attainable only through Christ. "There is no other way you can taste or have virtue, because he is the way and the truth, and whoever holds to this way and truth cannot be deceived."[37]

If Catherine of Siena personified the powerful, celebrated spiritual director, Julian of Norwich was an example of another class of spiritual director popular during the late medieval period, the anchorite. Uniquely medieval, anchorites embraced a solitary life in the midst of a community, usually by being enclosed in small huts attached to parish churches. They were totally dependent on the community for their material well-being, and in return they were expected to support the community's spiritual well-being. One window in the hut facilitated both functions; it was where food and the like was passed in and where spiritual advice was given out. Because "one who lives the life of sublimity"[38] was so isolated "from all the world's noise, so that nothing may prevent her from hearing God's voice,"[39] the community depended on anchorites to live, according to the *Ancrene Wisse,* "so holy a life that the whole of Holy Church, that is, Christian people, can lean upon them and trust them, while they hold [the Church] up with their holiness of life and their blessed prayers. This is why an anchoress is called an anchoress, and is anchored under a church like an anchor under the side of a ship, to hold that ship so that waves and storms do not overturn it." The people support the anchorite in return for the spiritual benefits, and "every anchoress has made this agreement."[40] Part of this agreement entailed the women being available to people who "come to them at the window"[41] for guidance, particularly about the eight crucial conditions of human life: "1) the shortness of this life; 2) the difficulty of our journey; 3) our goodness, which is so meager; 4) our sins, which are so many; 5) death that we are certain of, and uncertain when; 6) the stark judgment of Doomsday and its extreme strictness . . . ; 7) the grief of hell . . . ; and 8) how great is the reward in the joy of heaven world without end."[42]

Not all anchorites were women, but at least in England, where we have the most reliable numbers, the vast majority of them were.[43] Julian of Norwich was one such woman. We know little about her from contemporaries, and there is no evidence to indicate she was powerful, influential, or well known outside her region. Her social position, in other words, was diametrically the opposite of that of Catherine of Siena. Four wills from the era do mention bequests to an anchorite called Julian (named after the church of St. Julian in Conesford at Norwich to which she was attached). Margery Kempe, the unconventional housewife and mystic, not only verifies her quiet existence but also certifies Julian's local ministry of spiritual direction. Margery wanted "to find out if there was anything wrong with what she felt" about certain spiritual matters, so she and her husband went "to many other places, and spoke with God's servants, both anchorites and recluses," to discuss "her feelings and her contemplations to several of them."[44] Julian of Norwich was one of the spiritual advisors Margery sought, "for the anchoress was expert in such things and could give good advice."[45] To Margery's question of whether there was "any deceit on her experience" of "compunction, contrition, sweetness and devotion, compassion with holy meditation and high contemplation, and full many holy speeches and dalliance that Our Lord spoke to her soul," Julian answered by reassuring Margery of their validity. "Any creature that has these tokens" of right faith and right belief may believe her mystical experiences are from God. "I pray God grant you perseverance. Set all your trust in God and fear not the language of the world."[46] Margery records only this short advice of Julian's, but there was more, for "much was the holy dalliance that the anchoress and this creature [Margery] had by communing in the love and Our Lord Jesus Christ the many days they were together."[47] We do not have any other sources pertaining to Julian's guidance of individuals, but Julian did write a general treatise for spiritual direction, so we have a rather clear idea of her practice, method, and message.

Julian's first tenet of spiritual direction was her belief that the personal is the foundation for the universal. She was quite insistent upon this. Her *Book of Showings* appears at first glance to be an autobiographical explanation of her mystical experiences, but upon closer examination it is evident that she believed her personal experiences were intended to be used as vehicles to communicate her spirituality and direction. In both the long and short versions of *Showings*,[48] she included almost verbatim a statement of this principle: "Everything that I say about me I mean to apply to all my fellow Christians, for I am taught that this is what our Lord intends in this spiritual revelation. And therefore I pray you all for God's sake, and I counsel you for your own profit, that you disregard the wretch to whom it was shown, and that mightily, wisely and meekly you contemplate upon God, who out of his courteous love and his endless goodness was willing to show it generally, to the comfort of us all."[49] All should accept her revelations "as if Jesus had shown it to you as he did to me." The reason for this belief is as revealing as it is profound. "I write as the representative of my fellow Christians," because "in general I am in the unity of love with all my fellow Christians. For it is in this unity of love that the life consists of all men who will be saved."[50]

Julian's spirituality is deeply rooted in this paradoxical relationship between the personal and the communal. The mystic is by definition one who experiences God in a deeply personal and individualistic way, yet Julian insisted that her own mystical experience had a purpose beyond her personal salvation. It was meant for the community's salvation. "I am not good because of the revelation, but only if I love God better," Julian argued, "for it is common and general, just as we are all one; and I am sure that I saw it for the profit of many others."[51] Once the individual tries to find happiness alone, happiness eludes him or her. "If I pay special attention to myself, I am nothing at all; but in general I am, I hope, in the unity of love with all my fellow Christians."[52] Love for every single person is absolutely essential for happiness, for "if any man or woman withdraws his love from any of his fellow Christians, he does not love at all, because he has not love towards all."[53] Indeed, it is "truly love which moves me to tell" her directees about her mystical experiences "for I *want* God to be known and my fellow Christians to prosper, as I hope to prosper myself."[54] So important was this relationship between the personal and the communal that she summarized her beliefs yet again in the next chapter: "And in all this I was humbly moved in love towards my fellow Christians, that they might all see and know the same as I saw, for I wished it to be a comfort to them all, as it is to me; for this vision was shown for all men, and not for me alone."[55]

In the earlier, short text Julian addressed the question of her gender and the role of spiritual directors in general, making sure it was not an obstacle in the reception of revelation. "I know very well that what I am saying I have received by the revelation of him who is the sovereign teacher," so just "because I am a woman, ought I therefore to believe that I should not tell you of the goodness of God, when I saw at the same time that it is his will that it be known?"[56] The truth of the message is always more important than the messenger. All spiritual directors are ultimately only funnels for Jesus' direction, and "it pleases [God] that we seek him and honour him through intermediaries."[57] If the directee concentrates on the message, "then will you soon forget me who am a wretch, and do this, so that I am no hindrance to you, and you will contemplate Jesus, who is every man's teacher."[58] Given the fact that Julian was the earliest known woman author in English writing about spiritual experiences, and given that the religious was always intensely scrutinized by medieval society, her repeated insistence that she was only a conduit for God's message may have arisen in part to divert attention from her pioneering status, but not entirely. She insisted so vehemently because she believed it so sincerely. To Julian, the individual was complete and happy only when he or she lives, loves, and helps others be happy. "The more that I love in this way whilst I am here, the more I am like the joy that I shall have in heaven without end," Julian maintained.[59] Thus, to Julian, the individual reaches perfection only when she realizes the limits of her individuality, and the community is saved only when it accepts its dependence on the individual. It was a message that her fourteenth-century society needed to hear; it was an answer to many pertinent issues of the day.

Much attention has been given of late to Julian's portrayal of God, particularly Jesus, the second person of the Trinity, as mother. There are many theological traditions and doctrines to explain Julian's use of this imagery,[60] but not the least important reason why she chose this imagery was because of its effectiveness in spiritual direction. Like Hadewijch, Julian believed in the premier position of reason in the search for happiness. "Man endures in this life by three things," Julian acknowledged, and "the first is the use of man's natural reason."[61] Like Angela of Foligno, Julian understood that "we shall all come into our Lord" only by "knowing ourselves clearly."[62] Like Gertrude the Great, Julian reminded her audience that the proper exercise of free will was key in finding happiness; "any man or woman" who "voluntarily chooses God in his lifetime" will be so rewarded "with an endless love."[63] But Julian was frustrated in her ability to communicate certain aspects of her spirituality, because she "cannot show the spiritual visions as plainly and fully as I should wish."[64]

In the years between her first book of direction and her second, her frustration drove her to find images more capable of communicating the fullness of her message. Hence, the later text contains lengthy discussions of Jesus as mother that are absent from earlier text. Julian saw that one of the most pressing problems for those seeking eternal happiness through Christ is to trust that "he is here with us, leading us, and will be until he has brought us all to his bliss in heaven."[65] Although the search for happiness "is common to all," it is often elusive and is accompanied by "unreasonable depression and useless sorrow." Regardless of the difficulty, the soul must "surrender itself to him in true confidence," and "have great trust in him."[66] Julian employed the imagery of motherhood to communicate her spiritual thesis concerning the centrality of trust in the search for eternal happiness. Jesus "wants us to love him sweetly and trust in him meekly and greatly" and to believe firmly in his "gracious words: I protect you very safely."[67] For Julian, the figure of mother expressed these realities better than any other: "The mother's service is nearest, readiest, and surest: nearest because it is most natural, readiest because it is most loving, and surest because it is truest."[68] By portraying Jesus as mother Julian gave her directees, all of whom had personal, direct experience of maternal imagery, a more accessible way to comprehend her message. If Jesus is our mother, then we are his children and must act accordingly and trust accordingly. Jesus "wants us to show a child's characteristics, which always naturally trusts in its mother's love." In moments of doubt and difficulty "He then wants us to behave like a child," trusting reflexively in a mother's natural love. As a child, "when it is distressed and frightened, it runs quickly to its mother," so, too, must the soul in distress call upon Jesus, saying "My beloved Mother, have mercy on me. I have made myself filthy and unlike you, and I may not and cannot make it right except with your help and grace."[69] Jesus "wants us to trust that he is constantly with us,"[70] and Julian communicated this belief quite effectively—and affectively—through the imagery of motherhood.

The vita of the widowed anchorite Dorothea of Montau offers us additional information about late medieval women directors. The first native Prussian saint, Dorothea's life is testimony to the universal churchwide acceptance of women spiritual directors as unexceptional. Indeed, that she was an anchorite seemed more noteworthy to her society than her being a spiritual director. When Dorothea asked her confessor to help her obtain a anchorhold, "he did not hold out much hope for her being given one there, [so] Dorothea often thought of abandoning the place to search for another within her homeland or in a foreign country where she could be enclosed."[71] Apparently "at that time hermitages within churches or attached to them were uncommon in Prussia and not seen there."[72] Her functioning as her children's and the region's spiritual director, on the other hand, was not noted or of concern. The confessor she referred to here was Johannes von Marienwerder, who was also the author of her vita. We have neither writings of hers nor any sources on the content of her direction. That her male confessor provides the only information we have concerning her as spiritual director presents us with a different perspective than that found in the first-person accounts of Catherine of Siena and Julian of Norwich.[73] Johannes wrote the vita specially for Dorothea's canonization inquiry[74] and thus was intent upon making her appear in her most favorable light. He obviously had no concerns that the papal commission would object to her role as spiritual director, or he would not have emphasized them thus. He recorded that he was one of two confessors she had, to whom "the Lord drove her to be obedient to B."[75] At the same time he acknowledged the reciprocal nature of their spiritual direction, by referring to "her two confessors and spiritual sons B and P."[76] Apparently he thought women spiritual directors to be totally within the norm both in theory and in practice.

The first directees of Dorothea that Johannes identified were her nine children. Dorothea "labored diligently to give birth to these same children spiritually whom she previously had born into the world physically." Her every waking moment was dedicated to their spiritual training. She "rose early with great anxiety to pray" for each child individually, shedding many tears for them. From infancy she "taught them to fear God and avoid sin," and she continued to do so well into their adult years. "With words and deeds, good habits, and examples she enticed and guided her children toward a desire for the kingdom of heaven," all with more "diligence, concern, labor, and exertion" than it took "in giving them physical birth." During her married years she was "equally dedicated [in] her effort on behalf of other people whom she, according to God's exhortation, was to bear to him in spirit through fervent prayers and excellent guidance to a new, virtuous life. Weeping and praying, she diligently interceded with God for her spiritual children while punishing and admonishing them and doing what she could to draw them to God."[77]

Dorothea had quite definite ideas about the dimensions of spiritual direction. "'The human soul is to be trained with the greatest diligence,' the Lord told Dorothea,"[78] and Johannes repeats to us. It was Jesus Christ himself "who taught her how to teach her spiritual children." The first lesson was whether contrition for sin was complete. "The second lesson is this: Mark whether you flee from sin because of your sincere love of God and true fear and horror of evil or because of mere cowardice." The third lesson is learned from tears of compunction, and the fourth is the revelation of "the correct paths to eternal life."[79] The spiritual director "must practice self-denial, abandon himself, the world, and everything in it" if he or she wants to "train the soul so well that she will become ample and great."[80] Throughout the vita it is maintained that Dorothea "was instructed by the Lord," learning "self-discipline and self-castigation under the master who has his lectern in heaven and his school on this earth."[81] Such statements are common in women's spiritual works and often are considered to be rhetorical devices to direct attention from the women's authoritative voice and present them instead as a channel for God's voice. In Dorothea's vita, however, the statements reveal more: a distinction between the roles of confessor and spiritual director. Sometimes the insistence of Jesus as Dorothea's spiritual director is used to emphasize that Johannes functioned primarily as her confessor in sacramental confession, a ritual distinct from spiritual direction. Once when Dorothea "offended God" and "was tormented by the wounds of [her] sins," she searched and searched for her confessor. "I could not have returned to my hostelry without confessing first, no matter whether it was morning or night," she admitted, for until she found her confessor "I could not find rest or peace until I had gained [God's] assurance of my sins have been forgiven." It was the "most beloved Lord God," however, who "demonstrated to me just how fragile the soul is," and it was the Lord who told her to wash her soul "with gentle tears and with confession" to "improve yourself."[82]

In other places in the vita, however, the roles of confessor and spiritual director appear merged into one person, Johannes. After her husband died Dorothea wanted a more intense religious life, so she sought out the "advice of the man who had been recommended to her" in Marienwerder: Johannes.[83] There "she opened her heart" to Johannes "who from then on was her confessor, always revealing as much to him as the Lord inspired and ordered her to say." After some time Dorothea "committed herself to the authority of B, just as the Lord had commanded," in a relationship that God described to Dorothea thus: "You both shall often take to heart how I have brought you together. I have united you just as two people are bound to one another in marriage, and for this reason each of you shall take on the burdens of the other and one help the other so that you both may come to eternal life."[84] Rarely, if ever, has the reciprocal nature of spiritual direction been more poignantly stated, yet the relationship was identified as between confessor and confessee, not between director and directee.

This is not surprising, given the fact that the term "spiritual direction" does not become common until the early modern period. Still, it is most unfortunate for historians of spiritual direction, because the indiscriminate use of the terms "confessor" and "confessee" (or penitent) for relationships more accurately described as director and directee leads to confusion. It is in large part responsible for our failure to identify the pervasiveness of spiritual direction and of women spiritual directors. There are many other reasons, too, why spiritual direction was often swept under the cover of confession. Both confession and spiritual direction enjoyed increased popularity at this time. Sacramental confession had become more widespread as a result of the Fourth Lateran Council mandating annual participation, but post-Fourth Lateran confession adopted many characteristics first practiced in spiritual direction. The adoption of penitential handbooks to facilitate good confessions, for example, encouraged many people to examine their interior motives and intentions for the first time within the ritual of the sacrament. The goal of confession, according to Jean Gerson, the leading Parisian theologian of the day, in his *On the Art of Hearing Confessions,* is complete contrition and repentance. To help the penitent achieve that end, the model confessor must become "like the most learned physician of spiritual diseases" who knows everything there is to know about the patient and the disease, much like the model spiritual director.[85] Similarly, the purpose of spiritual direction is to guide an individual toward his or her spiritual end, happiness, by helping people avoid counterproductive behavior and encouraging productive behavior. Thus we can see that the line separating confession and spiritual direction often overlapped by the late Middle Ages. Both confession and spiritual direction shared the same goal, to bring one to happiness. Confession had one big advantage over spiritual direction, though, which may account for its overshadowing the latter during this period. It was mandatory and, therefore, provided a readily available opportunity to talk privately with a person supposedly trained and knowledgeable in all things spiritual. Such success to other possible spiritual teachers might not be as available for those outside a monastic or urban setting. Still, as we shall see, confession did not swallow up spiritual direction. It did become an important place where spiritual direction was given, but not to the degree that it was the sole place for direction. Holiness was still the chief characteristic that qualified people as directors, and directees knew this. They often looked, therefore, to those they perceived to be holy. This meant that holy women spiritual directors were in as much demand as ever, just as unholy confessors were not. Spiritual direction and confession remained distinct, even though the practice of both was increasingly performed by the confessor.

Johannes von Marienwerder, the source of our information concerning Dorothea, acknowledged this distinction himself. He (and the unidentified prior he referred to as P) was Dorothea's confessor, but throughout the vita he asserted that Dorothea herself and the Lord were the sources of her guidance; directives were always introduced with qualifiers such as "the Lord told her," or "the Lord provided Dorothea with a rule."[86] Johannes never claimed that he was the source of any significant direction. All guidance came from within Dorothea. "Dorothea studied and learned how to win a clean conscious," independent of her confessor, just as she learned to discern "the nature of her sins, their degree, their frequency, and where, when, with what intent, determination, and purpose she had committed them," and "to discern, see, hear, smell, taste, and feel the eternal treasures." Only after learning all that on her own did she approach her confessors that "they may weigh [her beliefs] and check them against holy writ and examine whether they are correct."[87] Thus, Dorothea's very interaction with her confessors was determined not by those confessors but by the spiritual guidance she received directly from God. "You shall put yourself entirely into your confessor's care," the Lord told Dorothea, according to Johannes. "If he wishes to speak to you, speak to him, and when he wants to give me to you, receive me with love."[88] When Johannes did offer judgment it ironically was still under her directions, as, for example, when she reported to Johannes that the Lord told her, "'Tell your confessor that he may state truthfully that you show perfect trust by not worrying in the least about temporal matters.'"[89]

Birgitta of Sweden's relationship with her confessor was remarkably similar to that of Dorothea's. Like Dorothea's vita, Birgitta's vita was written by her two confessors, Prior Peter and Master Peter, in later life specifically to support her canonization. Like Dorothea's confessor and hagiographer, they would not have included questionable behavior or roles of Birgitta in a written text that was meant to prove her orthodoxy to the hierarchy. The two Peters tell us about Master Matthias, Birgitta's first confessor, whom "she obeyed in all her difficulties," and yet they also write that Birgitta determined her spirituality independent of him. Matthias was twice called "a very expert and devout master of theology," a writer of "an excellent gloss on the whole Bible" and "many volumes of books,"[90] but mentions no spiritual qualifications to guide this extraordinary mystic. Because "I do not speak to you for your sake alone, but for the sake of the salvation of others," the Lord said to her during a mystical experience, it was necessary for her to have a hierarchical seal of orthodoxy in order to have her message easily accepted. Therefore, she was to "go to Master Matthias, your confessor, who has experience in discerning the two types of spirit" and "say to him on my behalf what I now say to you: you shall be my bride and my channel, and you shall hear and see spiritual things."[91] At the beginning of her revelations "she was at once instructed to obey that same Matthias," but at the very same time she was moved by the Spirit, not by her confessor, "to stay in a monastery of Cistercian monks."[92] It was Matthias who sought Birgitta's aid, not vice versa, when he needed specific spiritual guidance while being "stricken with a certain temptation."[93] To Birgitta confessors were there to keep her from self-deception and to provide access to sacramental forgiveness because "in her there was fear of God together with great love of him."[94] She scrupulously "noted that word by which she offended God, in order that she might not neglect to confess it and make satisfaction for it by means of penance."[95] The vita talks at length about the role of confession in Birgitta's life, reporting that in the early part of her life "she was accustomed to confess every Friday" and that as spiritual life progressed "she also confessed more than once on every day."[96] When "making her confession, she was very humble and very prompt in fulfilling whatever things were enjoined upon her."[97] Nowhere does the vita mention Birgitta seeking spiritual guidance in confession.

The vita is very clear, on the other hand, that Birgitta was a spiritual director. She was the spiritual guide of her husband who "occupied himself—

at his wife's advice and admonishment—in learning to read the Hours of the most Blessed Virgin Mary"[98]; she was Master Matthias' spiritual director. Throughout her life her fame as a spiritual director grew, and her method for direction became distinct enough for her hagiographer to report about it. The account also reveals Birgitta's use of confessors and is worth quoting at length.

> [W]hen any person asked her about some doubt in his conscience and sought from her advice, and a special remedy that would be very good, she then used to answer him: "Pray to God about this. And we too shall think, and we shall do what we can for you—although I am an unworthy sinner." In fact, after three days or so and sometimes on the very same day, she would answer that same person. . . . If, in fact, she was ill, she called her confessor and her writer—a secretary specially assigned to this—and then, with great devotion and fear of God and sometimes with tears, she reported to him those words in her own vernacular. . . . And then the confessor said these words in the Latin tongue for the writer, and he wrote them down right there in her presence. And afterwards, when the words had been written out, she wanted to listen to them; and she listened very diligently, and attentively. And so she gave or sent this writing to those who were making the inquiry. This has often—yes, very often—been proven in experience by the lady queen and the archbishop of Naples; also by the queen and the king and the prince and many others from the kingdom of Cyprus and from the kingdom of Sicily; and by men, and by women too, from Italy, from Sweden, and even from Spain.[99]

To Birgitta of Sweden's own confessors, then, she was the director and they were her facilitators. No one considered any of this unacceptable or extraordinary.

The tradition of women's spiritual direction becomes more extensive when we reexamine some spiritual treatises and identify them more specifically according to purpose: Many are treatises of spiritual direction. Catherine of Bologna's *Seven Weapons of the Spirit* is such an example. During her own lifetime Catherine was well respected in her narrow religious circle, but not beyond it. Her direction was specifically for her fellow religious. Her ministry of spiritual direction probably began while she was mistress of novices in the Ferrara monastery of Corpus Domini, and it continued during her abbacy at Corpus Domini Monastery in Bologna. That Catherine placed great emphasis on the Word can be seen in numerous paintings she created. Illuminata Bembo, Catherine's close associate and hagiographer, reported that "Gladly, in the books and in many places of the monastery of Ferrara [Catherine] painted the Divine Word."[100] In her famous *Redeemer* painting Catherine visualized Him as the Incarnate Word and placed an inscribed book over Christ with the message "In me omnis gratis. In me omnes (spes) vite et veritatis. In me omnis spes vite et virtutus,"[101] thus communicating through this imagery her belief that salvation comes through reading words about the Word.

This was her approach to spiritual direction in general. She promoted reading as a means to learn life's mysteries. To that end she established an extensive, impressive library for the women in her Bologna monastery,[102] illuminated a breviary (also filled with images of the Incarnate Word), and recorded her principles for spiritual direction in written form. In an age that gave birth to the printing press and the new era in human communication it ushered in, Catherine actively promoted the written word and maintained that its potential in spiritual direction was as great as that of example. *Seven Weapons of the Spirit* "was composed with divine assistance by me, a little barking puppy," Catherine explained, to direct others toward perfection. Catherine also possessed the same strong sense of obligation we saw in many other spiritual directors. "I, the above mentioned little puppy, am writing it with my own hand, only because I fear divine rebuke if I fail to say what may help others,"[103] particularly those women religious placed in her spiritual charge. "I shall write here below, after a while, some instructions to comfort those persons who have entered this most noble battle of obedience, and find themselves strongly opposed and troubled by their own will, by their desires and opinion," Catherine pledged to her sisters. It was her duty, after all, "to show the truth of this as I proceed further." She proposed to the soul who wished to find perfection "to serve God in the spirit of truth" that it "should first of all cleanse its conscience by a pure and general confession."[104] After that was done, spiritual direction could effectively begin.

One of the first weapons Catherine prescribed was the unreliability of oneself and the consequent importance of spiritual direction, for how "can a simple nun, especially one who has just entered the religious life, be arrogant enough to prefer her own thoughts and zeal? She should much rather accept the advice and will of her prioress or mistress."[105] Consistent with her overall spirituality, Catherine advocated the word of Scripture as the most powerful weapon and the weapon with which one tests and judges the goodness or badness of one's spiritual insights. From Scripture "we should derive counsel, as from a most devoted mother, in all the things that we must do; as we read of that prudent and holy virgin Saint Cecilia, where it reads, 'she always carried the gospel of Christ hidden in her breast.'" The Word contained salvation, "so, beloved sisters, do not allow those daily lessons read in choir and at table to be in vain [but] think that the epistles and gospels you hear every day at Mass are fresh letters sent to you from your heavenly spouse."[106] In this way the happiness of heaven, with all "the joys that are prepared for us there," will be enjoyed.[107]

Similar attitudes toward reading, confession, and the Word are found in the spiritual direction of Clara Gamabacorta. Clara possessed most of the characteristics historians Donald Weinstein and Rudolph Bell identify as typical of a fourteenth-century saint. Although she attained only the rank of blessed, she lived during the era of Italian urban saints who "served their towns as welfare agents, counselors, and peacemakers;" historian Richard Kiekhefer says Clara functioned in all these roles.[108] As with all her predecessors, society viewed her being a spiritual director as completely within the norm. Because of her emphasis on reading as a vehicle for spiritual direction, her guidance, like that of Catherine of Bologna, is a precursor of the future. In correspondence with one of her directees, famed merchant Datini of Prato, Clara urged him to "be faithful to spiritual reading and to prayer; make your confessions with great care, for confession maintains the soul in the state of purity and prepares it to receive Divine grace."[109] To Datini's wife Marguerite, Clara elaborated on her reasons for emphasizing reading. Twice she told Marguerite that "it gives me great pleasure to learn that you know how to read, and I pray you to make use of your knowledge."[110] In the first letter she identified three advantages to be gained by spiritual reading. "The Saints have taken pains to write books, that we may see ourselves therein, that we may adorn ourselves in virtue, and that we may remove the stains of sin."[111] Clara then discussed prayer and love of God but returned to reading and confession before ending her direction. "Since you know how to read, partake of this holy food and especially grow strong in holy charity," she urged. "Go often to confession, so that on the holy day of Christmas you may receive the adorable Incarnate Word."[112] In a letter written a year later she provided different reasons for turning to reading for direction. "Saint Augustin says reading prepares the soul for prayer. The soul is filled by reading, and in praying, by means of this reading, the soul receives in itself great light; and, between prayer and reading we are instructed by Jesus Christ and His saints on what we have to do to acquire grace in this life, and glory in the next."[113] In a letter addressed to both husband and wife the importance of reading and the Word were again emphasized: "Nourish yourself by holy reading," Clara counseled, for "the Eternal Word . . . was sent to teach us."[114] We also see in this letter that Clara lived what she preached; she begged them "to buy us a quantity of paper for we are writing a book of Epistles, and we are in great need of it. It would be a very great charity as we are poor in books, and this one is very necessary; we shall write it ourselves."[115] The importance of reading in the search for meaning was so pivotal that she apparently begged books from strangers. She once heard that a certain hotelkeeper named Angela had "relations with many religious, and that books often come into your hands," so she approached him for some books. "I beg you to let me know if you have on hand 'a Book of Lessons or a Bible, for we are very poor in books, and, moreover, we must have these for the recitation of the Divine Office." Since it cost too much to have them written for the community, "we cannot even think of it, but if some already written should fall into our hands, I trust in God and in some good persons to see that they are paid." She ended with a final confession that "we like all books, but we have named the above two, because we are in greater need of them."[116]

Chapter 6

$\backsim\!\!\!\!\!\bigcirc\!\!\!\!\!\sim$

Confessors, Spiritual Directors, and Women

During the sixteenth century Reformation leaders called into question almost every aspect of Christian practice, demanding its reform, its reformulation, its justification, or even its demise. Sacramental confession was a primary target, and the frontal attack reformers made on it in effect encouraged the union of the roles of the confessor and the spiritual director that had begun during the late medieval period. Throughout the ages Christians have used different rituals to forgive sinners their sins, but common to them all are two basic goals, the preservation of external social order and the rectification of interior moral order. Reformers objected to an overemphasis on the former to the detriment of the latter.[1] The appearance of the confessor-director at this moment in history was indeed fortuitous, for the new composite figure was able to establish a more satisfactory balance between the two goals. The success of this readjustment can be seen in the popularity of the confessor-director in the Counter Reformation church. Confession became more frequent since the Fourth Lateran Council had mandated annual confession and less onerous because it included much-desired spiritual direction. Francis de Sales, a proponent of a new affective confessional method, advised laity to "communicate more often" than once a month, "but as to confession, I advise you to go even more frequently."[2] The laity responded with enthusiasm for a variety of reasons. Because advocates like Francis de Sales believed that the relationship between the confessor-director and the confessee-directee should be personal and unique, a new freedom, even an obligation, to seek out various people for guidance until the proper one was found was established. As he wrote to Angélique Arnauld, reform abbess of Port-Royal, "there is no harm in searching from several flowers the honey one could not find upon one alone."[3]

The development of the confessor-director figure had a particular impact on women. The emphasis on compatibility between confessor-director and directee gave women a good deal of leverage in a penitential system where they previously had none. Instead of submitting to the demands of one's parish priest or to those of an appointed cleric, as mandated by the Fourth Lateran Council, confessees were now in a position to exercise control over the confessor by choosing the confessor. Women religious were quick to realize this was an advantageous innovation and often fought hard to retain it. Teresa of Avila correctly identified it as especially crucial to women. "I am sure the relations of penitent and confessor, and the type of confessor to be chosen, are very important matters, especially for women," Teresa argued, because "the Lord gives these favours [of spiritual life] far more to women than to men" and women make "much more progress on this road than men."[4] The issue was so essential to Teresan reformers that a papal brief written soon after her death firmly asserted the right of Carmelite women to choose their own confessors, although that right remained a dividing issue within the male and female branches of the Order throughout the first half of the seventeenth century.[5] We need wait only one generation to see how extensive were the implications of such choice; Mary Ward, founder of the Institute of the Blessed Virgin Mary, argued that women should use this freedom to choose only those confessor-directors who believed in the equality of women before God. When a priest cynically observed that Mary's recently established Institute would fail after the initial enthusiasm died, because "when all is done they are but women," Mary angrily expounded her views on women's nature at length. "It is true fervor does many times grow cold, but what is the cause? Is it because we are women? No, but because we are imperfect women. There is no such difference between man and woman," Mary began. "Fervor is not placed in feelings but in a will to do well, which women may have as well as men. There is no such difference between men and women that women may not do great things." Women must choose only confessor-directors who acknowledge this central fact if their relationship is to be fruitful. "That you may not be deceived, you may know [confessor-directors] by the fruits of their counsel. Those by whom you have been directed have generally been the best directors of all. For what can this profit you, to tell you you are but women, weak and able to do nothing; and that fervor will decay? I say what does this profit you but bring you to dejection and without hope of perfection." Since "all are not of this opinion," women must search for confessor-directors who hold proper opinions concerning women's spiritual nature.[6] Mary Ward did not look to the confessors for proper spiritual assessment. Rather, confident that her own opinion was correct, she demanded that clerics endorse hers. Mary's certainty in this matter even leads to a trifle smugness at times, as we can detect in her reaction to an unenlightened cleric. "There was a Father that lately came into England whom I heard say that he would not for a thousand of worlds be a woman, because he thought a woman could not apprehend God. I answered him nothing, but only smiled, although I could have answered him by the experience I have of the contrary.[7]

If only women were claiming freedom of choice for confessors, we might suspect the newly discovered freedom was a disguised effort to sabotage the power of the confessor-director, but perhaps the most vocal champion of the freedom was John of the Cross. That Teresa was John's spiritual mother might explain how he came to realize the importance of this freedom, but it does not nullify his arguments or the historical influence he exerted over all future mentors and students of spirituality. Large portions of John's *Ascent of Mount Carmel* and *Living Flame of Love* are devoted to discussion of spiritual direction, the harm bad directors can do, and the subsequent necessity of freedom to chose one's own confessor.[8] In *Ascent of Mount Carmel* John writes thus: "Some spiritual directors are likely to be a hindrance and harm rather than a help to these souls that journey on this road. Such directors have neither enlightenment nor experience of these ways. They are like the builders of the tower of Babel. (Gen 11:1–9) When these builders were supposed to provide the proper materials for the project, they brought entirely different supplies, because they failed to understand the language."[9] In *Living Flame of Love* John feels compelled to repeat his judgement with even more stinging condemnation of inadequate confessor-directors: "Many spiritual directors cause great harm to a number of souls, because, in not understanding the ways and properties of the spirit . . . they instruct them in other baser ways, serviceable only to beginners, which they themselves have used or read of somewhere. Knowing no more than what pertains to beginners—and please God they would even know this much—they do not wish to permit souls to pass beyond these beginnings."[10] Because of this very real danger, John of the Cross emphasizes the need for freedom in choosing the confessor-director: "It is very important that a person desiring to advance in recollection and perfection, take care into whose hands he entrusts himself, for the disciple will become like the master, and as is the father so will be the son. Let him realize that for this journey, especially its most sublime pasts (and even for the intermediate parts), he will hardly find a guide accomplished as to all his needs."[11]

The essence of John's opinion was summarized in Jean Surin's concise comment that confessors "must be chosen in good faith and by an entirely free choice and not by force."[12] The goal is proper direction, not a test in obedience. As Teresa of Avila's dear friend Ana de San Bartolomé reported about Carmelite Superior General Tomas de Jesus, "The Superior General desires that if a confessor does not give comfort, he should be changed."[13] Domingo Báñez considered women's choice of confessor not only to be essential but also evidence of her spiritual state. In fact, Báñez considered the spiritual gains made under the direction of a confessor ultimately to be the result of the chooser rather than the chosen. In his formal review of Teresa of Avila's *Life* for the Vallodolid inquisitors, Báñez concluded that Teresa's "great experience, as well as her humility and discretion in always choosing confessors with illumination and learning," was responsible for her spiritual well-being.[14]

This freedom of choice led in unexpected directions, away from a permanent wedding of the confessor and the director into one clerical person. First came the realization that while any cleric could theoretically be a confessor, only certain clergy could be good spiritual directors. One's choice of confessor-director, therefore, really hinged more on qualifications a candidate had as a director than as a confessor. Teresa of Avila reflected upon the matter in *Life* and concluded that three elements are necessary. First, "he must be a man of experience, or he will make a great many mistakes and lead souls along without understanding them or without allowing them to learn to understand themselves." Unless confessor-directors have direct, personal experience in spiritual matters, they will "afflict their penitents both in soul and in body and prevent them from making progress."[15] Second, "at the same time, this matter of self-knowledge must never be neglected. No soul on this road is such a giant that it does not often need to become a child at the breast again. (This must never be forgotten: I may repeat it again and again, for it is of great importance.)"[16] Unless the confessor-director possessed self-knowledge, confusion rather than progress would result. Teresa's last characteristic was knowledge. "It is of great importance then, that the director should be a prudent man—of sound understanding, I mean—and also an experienced one: if he is a learned man as well, that is a very great advantage. But if all these three qualities cannot be found in the same man, the first two are the more important."[17] She based this prioritizing on the belief that "learning is of little benefit to beginners," yet, lest she be misunderstood, quickly added, "I want to explain myself further." A director without learning should still be avoided, for "not being enlightened himself, he cannot enlighten others." Some may argue that enlightenment through learning is not necessary for perfection. Teresa agreed, but with an important qualification: "[A]lthough learning may not seem necessary for this, my opinion has always been, and always will be, that every Christian should try to consult some learned person, if he can, and the more learned this person, the better."[18] Teresa's main concern was that a confessor-director may be chosen solely on the basis of learning, whereas learning in the presence of spiritual experience or self-knowledge was much more beneficial. "I have already said that a spiritual director is necessary, but if he has no learning it is a great inconvenience. It will help us very much to consult learned men, provided they are virtuous." Teresa summarized her discussion by arguing that a person "will be making a great mistake if he chooses a director without these three qualities"; in fact, "that will be no light cross for the penitent to bear."[19] We see how important Teresa considered these characteristics in her reflections on her own spiritual directors. In the beginning of her spiritual journey one of her confessor-directors was Pedro Ibañez, who "was a very learned man, so that I knew I could rely on anything he might say to me." Exposure to Teresa's spiritual life eventually made him more contemplative and finally led him to retire "to a monastery of his Order" until he was commanded to once again return to Avila. "When he came back, his soul had made such progress and his spiritual growth had been so great that he told me after his return that he would not have missed going for anything. And I too could say that same thing; for previously he had been reassuring and comforting me only by his learning, whereas now he did so as well by the ample spiritual experience which he had acquired of things supernatural."[20]

Reflection upon the qualities necessary for good direction was premised on the belief in the necessity of direction when pursuing perfection. "In all this we need experience and a director; for, when the soul has reached this stage, many things will occur which it will need to discuss with someone," Teresa concluded in her last chapter of *Life*.[21] This was a lesson that Teresa learned personally. When she had trouble comprehending a particular spiritual experience, she found some relief when she sought Peter of Alcantara's direction and he responded with comprehension. "And that was all I needed; for I did not understand myself then as I do now, and I could not describe what I was experiencing. Since that time God has granted me the ability to understand . . . but just then I needed someone who had gone through it all himself, for such a person alone could understand me and interpret my experiences. He enlightened me wonderfully about them."[22] Accordingly, that a person has a spiritual director and that a spiritual director possesses certain qualities were ultimately more important than the religious status or sex of the director. Teresa herself declared that the most important director she had was not a cleric but a married man, Don Francisco de Salcedo, whom she turned to when she first tried to begin "another and a new life," that is, "the life I have been living since I began to expound these matters concerning prayer." She soon realized that she "did not understand the matter properly" and, therefore, needed "to seek diligently after spiritual persons with whom to discuss this."[23] She found this director in Don Francisco, who "directs all he does to the great good of the souls with whom he holds converse." So crucial was his direction to Teresa's spiritual development that "he seems to me to have been the beginning of my soul's salvation."[24] That he was not a cleric seemed of little consequence to her, as her one parenthetical remark concerning his status indicates; she tells us she had given him "the best general account of my life and sins that I could (not in confession, as he was a layman, but I made it very clear to him how wicked I was)."[25]

Likewise, when Jeanne-Charlotte de Bréchard complained to Jane Francis de Chantal that the women in her convent were having a difficult time finding a director who possessed the necessary qualities, Jane replied thus: "Since your sisters are not finding outside the monastery what their hearts desire, dear Sister, may they be satisfied with direction from you. Everywhere we find that it is best to be satisfied with direction from within the house."[26] While Jane repeatedly manifested utmost respect for confessor-directors during her life,[27] she still had no qualms bypassing them if better direction could be obtained elsewhere. That is why she has no problem with her own role as spiritual director to clerical confessors themselves. Besides the mutual director-director relationship that existed between Francis de Sales and Jane, she also was the spiritual director of other clergy. Her brother André, Archbishop of Bourges, openly turned to her for spiritual guidance, and their correspondence reveals how extensive this direction was. She meticulously regulated "activities of the day, spiritual as well as temporal," and even instructed him on the words and timing of his prayers: "Sometimes be content to stay ever so short a while in His divine presence, faithfully and humbly, like a child before his father. . . . You may if you wish, say a few words on this subject, but very quietly: 'You are my Father and my God from whom I expect all my happiness.' A few minutes later (for you must always wait a little to hear what God will say to your heart): 'I am Your child, all Yours; good children think only of pleasing their father.'"[28] When Noël Brulart, a well-connected political figure in Henry IV's court, underwent conversion to an intense spiritual life, it was Vincent de Paul who suggested he petition Jane to become his spiritual director. She took on the task when Commandeur Brulart was just beginning his spiritual journal, guided him as he entered the Oratory and was ordained a priest, and remained his spiritual advisor as he retired "to spend the rest of [his] days in great serenity and true devotion."[29] Her direction of clerics was, of course, in addition to the spiritual guidance she gave to her own growing communities, her family members, and laymen and laywomen acquaintances.

Ironically, the advent of the confessor-director may have inadvertently given women more freedom to direct. Sincere confessor-directors knew how essential experience, self-knowledge, and learning were to direction and actively encouraged their confessees to seek direction from others, female as well as male. Teresa of Avila tells us that at one point in her life when she "saw that no one understood me," her confessor "gave me permission to relieve my mind by talking to [Doña Guiomar de Ulloa] about certain things, because for a multitude of reasons she was a suitable person for such confidences." Indeed, Teresa tells us her spiritual director was "a person of very great intelligence," enlightened about "matters of which learned men were ignorant."[30] Most also had enough sense to look to others more experienced in spiritual direction for help when they encountered situations beyond their comprehension. Teresa of Avila's works are filled with references to confessors who turned to her for such help. When a nun in a Cistercian monastery[31] started to frequently "fall to the ground and remain there for eight or nine hours," her confessor did not know how to guide the woman. In frustration and desirous of guidance himself, "the confessor, who was a great friend of mine, came to tell me about it. I gave him my opinion—that she was wasting her time, for these fits could not possibly be raptures or anything else but the result of weakness. I told him he must forbid her fasting and discipline, and provide her with some distraction."[32]

Teresa also tells us about a visit "by a confessor who was thought a great deal of, and who had been told by someone in confession that Our Lady often came to her, sat down on her bed and stayed talking with her for more than an hour." One or two aspects of the conversation were true, but "many of these things were foolish." As the confessor told Teresa this story, "I saw at once what was the matter." Teresa patiently explained to the confessor "that it would be best for him to wait and see if the prophecies came true, to make enquiries about other effects produced by the vision and to find out what kind of life that person was living."

The necessity these kinds of elementary inquiries seem apparent to us now, but to many sixteenth-century confessors not familiar with the tradition of spiritual direction they were anything but evident. "Not many years ago—quite recently, indeed—some most learned and spiritual men were greatly bewildered by things of this kind which someone reported to them," Teresa related. The only way to solve the dilemma was to do what this particular confessor did: "Eventually the man in question went and discussed them with a person [Teresa] who had had experience of favours granted by the Lord" and, in turn, was able to help his directee.[33]

Louise de Marillac did not even wait for Vincent de Paul to ask her for help; she offered it before he was even aware of the need. "I think much good would be done for our sisters if you took the trouble to visit them at the House," Louise openly directed the director. "Sister Hellot could be told of the benefits to the Company when the sisters become accustomed to being submissive to one another, and it could be made known that those who seem to have some authority should serve as an example."[34]

Louise apparently perceived herself to be Vincent's partner in directing and addressed him accordingly. "Monsieur, I have a most humble request to make of your charity," Louise wrote Vincent when circumstances prohibited her return to a retreat. "Please be so good as to visit our five sisters whom I left in retreat without having helped them much. I had led them to believe that I would return either this evening or tomorrow morning." Not able to complete the retreat, Louise was confident that Vincent could and would complete the spiritual direction she had begun.[35]

This was not a phenomenon unique to religious founders. We have correspondence between one confessor and Madeleine de Saint-Joseph, the first French prioress of the Parisian Carmelite monastery and confidante of Cardinal Bérulle.[36] The confessor wrote Madeleine for advice about problems that arose in his ministry of spiritual direction and about spiritual direction in general. "Concerning what you ask me, sir, whether we should give souls as much time as they wish, I will tell you that I do not agree with that." Her reason is based on what she herself learned from a woman more experienced in direction. "Our blessed sister Marie of the Incarnation told me on several occasions that it was sufficient to talk a quarter of an hour with souls and that anything beyond that was useless because, as she said, a quarter of an hour is enough to learn their needs and to respond to them."[37] Madeleine also did not hesitate to question the validity of confessors providing spiritual direction whose only qualification was that they were clergy: "It is very dangerous to meddle in it. One must be constrained and called to it by God." Many not truly fitted to the task "when they undertake the care of a soul, totally turn it upside down in order to form it in their mode." Better, Madeleine cautioned the cleric, "we should simply follow what we see God has done for it, since there is no question of leading everyone along the same path."[38]

Likewise, Louise de Marillac did not hesitate to direct spiritual directors. "Concerned about the spiritual weakness of another sister," Louise wrote to Guy Lasnier, abbot of Saint-Étienne-de-Vaux and advisor to one of Louise's houses for the Daughters of Charity in Angers. She asked him to pay keener attention to a particular problem, for it was widespread. "I fear, Monsieur, that the sisters are becoming overly sensitive to their interior feelings and that they are too used to introspection. I beg you most humbly, Monsieur, to take the trouble to look into this and into the ways in which it can be avoided."[39] Neither was Louise shy about giving instructions to Vincent de Paul about how he should direct her sisters in their ministry of spiritual direction: "Remember to advise the Ladies to be careful, while giving instructions, not to talk a lot to the seriously ill, even if they have not made a general confession. They should advise the patients to confess only sins they may have previously forgotten or held back, if they remember any."[40] When she did not have time to explain her spiritual decisions, Louise simply asked Vincent de Paul to trust her. "I believe it essential to remove this poor young woman early today," Louise hastily wrote to Vincent, "for several reasons which I will explain to your Charity when God gives me the grace to be able to speak to you."[41] Teresa of Avila's spiritual direction of her directors is well attested to, including one of her earliest confessors.[42] When she chanced to meet him years later, "the desire came to me to know about the state of his soul," so she debated whether to go and ask him. Three times she changed her mind until "finally my good angel got the better of my evil angel and I went to ask for him and he came to one of the confessionals to speak to me. I began to question him about his past life, and he to question me about mine, for we had not seen one another for many years."[43] When discussing the direction of Pedro Ibáñez she concluded that "he reassured me a great deal and I think [Teresa's own experience] was a help to him too."[44]

The life of Angela Maria Mellini presents us with a rather dramatic example of a directee turned director of her confessor. The relationship was denounced to the General Inquisitor in Bologna at the end of the seventeenth century by another directee of the confessor, jealous of the time the two spent together. A trial lasting some five months examined the matter in detail, and the records have survived. After a period of dissatisfaction with a confessor, Angela became convinced Jesus wanted her to have spiritual direction "more to his desire"[45] and found it with Giovanni Battista Ruggieri. The change was helpful, and under Ruggieri's request Angela began a spiritual diary that, among other things, reveals similarities between Angela's spirituality and that described by Catherine of Siena and Catherine de' Vigri of Bologna. Before long, the relationship took a new turn, as Angela realized her confessor-director was in serious spiritual trouble. She intuited the nature of the problem to be temptations against chastity, and she confronted him. After initial shock Ruggieri submitted to her direction; she mandated that he recite certain prayers with open arms every night, "so that the enemy does not tempt him to offend him with that sin."[46] Both Ruggieri and Angela testified before the Inquisitor that at the end of every confession first he would extend the penitential benediction over her, and then she would do the same over him. "She kept her hand raised moving it in the manner of the priests when they give the benediction," Ruggieri reported, "And I saw this through the holes of the confessional grate and I answered her 'let it be so' or else 'let it be done' or 'amen:' and if I did not always say it with my mouth I said it with my heart."[47] Angela agreed with this account, adding only that it was "my belief that it was the Blessed Virgin who blessed him, and not me who proffered the words for obedience to God."[48] Apparently neither the rituals nor the reciprocal confessor-director relationship between the two was considered to be wrong by the inquisitional office, for Ruggieri was reprimanded only for having confided in a woman about matters concerning chastity.

The rise of the confessor-director did not diminish the traditional role of the abbess and prioress (or mother superior, as the new active congregations called their equivalent leaders) as spiritual director of the women in her charge. Most of Jane Frances de Chantal's voluminous letters (over two thousand are extant) concern themselves with spiritual direction, and many were written to advise other religious superiors about their spiritual direction of women in their respective houses. The religious superior's primary role was to direct the woman toward their spiritual goal, and "this is why, as I'm always telling you," Jane wrote to Mother Marie-Jacqueline Favre, "you must free yourself from too much activity so that, insofar as you duties allow, you can be with your Sisters whenever they are together, to instruct them and encourage them."[49] The good superior must pray for "the spirit that is proper to spiritual mothers who, with a tender and cordial love, see to the advancement of souls"[50]; she must "be kind, sincere, trusting, open and communicative with the Sisters" and "win them through kindness, patience and instruction," Jane wrote to Mother Anne-Marie Rosset when she was at a loss as to how to direct a particular novice. "If [the novice's reform] is not clear and proven after more than a year, then in no way must there be any bargaining or talk of profession," Jane advised.[51] "I beg His divine goodness to give you the spirit that is proper to spiritual mothers who, with a tender and cordial love, see to the advancement of souls and who are never overeager," Jane wrote to another superior, Mother Marie-Madeleine de Mouxy. "That's the disposition I want for you, dearest; I can assure you that it will bring you many blessings. If it were necessary, I would even swear to you that if you do what I have just recommended, you will have an abundance of everything."[52]

One novice mistress Jane advised to "move on to a total surrender of yourself into the hands of our good God, so that, insofar as you can, you may help your own dear soul and those you are guiding, to be free of all that is not God"[53]; another novice mistress she just implored, "I beg you, dearest, follow [my] method."[54] Teresa of Avila believed the function prioresses fulfilled to be so pivotal in the spiritual direction of her nuns that she felt obligated to interrupt her narration of Carmelite foundations "to give some counsels by which prioresses may learn how to guide their subjects to the greater profit of their souls."[55] Similarly, Jane Frances tried to persuade a reluctant member of the Visitation community at Moulins that her prioress could direct her to happiness. "Didn't I promise you, my very dear Sister, that you would find in her [Mother Jeanne Charlotte de Bréchard] a person who would completely satisfy and console you? I ask God to grant that through this dear Mother you may receive the strength to grow in divine love. I advise you, in fact, I urge you, my dearest daughter, to be very open with her so that she may, by understanding your heart's weaknesses, bring to it appropriate remedies. If you do this, you will receive a great deal of satisfaction . . . and consolation, besides your spiritual benefit which you should seek above all else."[56] After all, every nun, Jane reminded the novices in Bourges, must accept "the will and judgment of your superior, whom you must allow to direct with you without any resistance."[57]

Confessor-directors, too, often acknowledged the advantages of mother superiors directing their nuns. María Vela tells us that her first confessor-director, Gaspar de Avila, believed that it was "a good thing to have my aunt [María's superior] . . . act for him" in matters of spiritual guidance. Thus, her aunt "gave permission" to María to receive Communion and "exercised the same authority in respect to prayer and penance that she did in respect to Communion." So complete was the aunt's spiritual authority that María says "neither the Prelate nor my confessor dared order that I do more than my aunt wished, lest she be distressed."[58] Father Roger Lee, confessor to Mary Ward's convent at St. Omer in Flanders, began by questioning the wisdom of wedding direction to the confessional. He argued that a person would only enter a confessional "to receive absolution of your sins, and not to allow it a place to tell histories." If a nun should "require further time of expostulation or of declaration, take other time for it out of the sacrament." This arrangement, however, sometimes resulted in some women being "more troublesome to their ghostly Fathers than I either wished or expected they would," so Father Lee decided the best course was to place direction back in the hands of the mother superiors: "And therefore, as far as my authority may sway with you, I command and request you in the boiling Blood of our Blessed Saviour, that you be careful on this point, permitting yourselves for the rest to be guided and governed by your Superior as by the lieutenant of God Almighty, as indeed she is, by whom hath bound Himself to concur."[59]

Teresa also discussed additional functions the prioress assumed with the advent of the confessor-director, that of overseer and intermediary. When a confessor misdiagnosed a spiritual condition and caused "a great deal" of suffering, Teresa warned thus: "[I]t is very important that each sister should discuss her method of prayer quite frankly with her prioress, who in turn must be most careful to keep in mind the temperament of that sister and her progress in perfection, so that she may help the confessor to understand her case better, and may also choose a special confessor if the one the sister has is not capable of treating such cases."[60] This role is even more evident in another incident involving "a nun and a lay sister" who experienced "vehement impulses of desire for the Lord" that were calmed only when they communicated daily. When their confessors, "one of whom was a very spiritual man," agreed with the women's demand, however, the situation only got worse: "The yearnings of one of them became so strong that she had to com-municate early each morning in order, as she thought, to sustain her life." The prioress realized something was awry and "wrote to me about what was happening, saying that she could do nothing with them," Teresa relates. "I realized at once what was happening, but until I got there I said nothing, for fear I might be mistaken.

> One of the confessors was so humble that as soon as I arrived and spoke to him, he believed me. The other was less spiritual, and indeed by comparison with the first had hardly any spirituality at all: to persuade this latter was quite impossible. I did not mind this, however, as I was in no way bound to him. I started to talk to the nuns and to give them many reasons, sufficient in my opinion, to prove to them that the idea that they would die without this par-ticular help was pure imagination. But the notion was so deeply rooted in their minds that no argument could eradicate it, and it was useless to reason with them further. . . . I was very stern with them, for the more I realized how refractory they were being to obedience, because they thought they could not help themselves, the more convinced I was that it was a temptation. They spent the first day in great suffering; the next tried them less; and in this way they gradually got better. . . . [61]

We can also see in this tale that while the confessor now possessed new significance as director, he was in no way above reproach or judgement. Teresa unhesitantly ignored the stance of the one confessor, because she deemed herself more spiritually advanced. If she perceived a confessor to be worthy of obedience, she willingly obeyed. If, however, he appeared unwor-thy, he was ignored or his opinion was overridden.[62] Ana de San Bartolomé tells us of such an incident in her life. Ana had what she called abundant "natural impulses to aid other souls," but her confessor was uneasy about them. "When my confessor saw that my zeal and love for other souls was lasting such a long time, he told me one day, 'Beware, my child, for this charity is of the Devil, and he is trying to deceive you.'" Ana was not satis-fied with his verdict, so she approached Teresa, then her superior, "to ask her if this were true, and I told her all that had happened. And she told me not to worry, that it was not the Devil, for she had gone through that same way of prayer, with confessors who did not understand it. With that I was com-forted and I believed that just as the Saint told me, it was of God."[63] Ana presents Teresa's contradiction of the confessor as unexceptional and feels no need to justify her own acceptance of Teresa's verdict over the confessor's.

When Ana moved to Paris to head a new Carmelite house, she again challenged the validity of her confessor's decisions. In this instance the con-fessor was the powerful Cardinal Bérulle, her ecclesiastical superior in France and chief patron of the Carmelites there. In other words, besides being her confessor, he was essential to the success of the Carmelite reform. Still she did not hesitate to challenge him. The conflict began when Bérulle began "disputing points in the Constitution and the Rule (of the Carmelite Order) about some things he wanted to change. I contradicted him, and he said he knew these things quite as well as I. I told him that was not so; that he must be great in book-learning, but that he had no experience, as I did, of matters concerning the Order, and that I would never agree to it." Ana would not budge from her stance at all, despite the fact that the confrontation went on for so long "that when I came away I was so ill they had to bleed me." And while she does not go into details about her confrontations with him within the confessional, she lets us know her rebellion against his decisions continued there: "And what most afflicted me was confession, for I had to confess with this man, while we were quite opposed because of those things which I observed going against the Order."[64] It should come as no surprise, given this confrontation, to learn that Ana considered the issue of choice of confessors to be central to the survival of the Carmelite reform. She counseled patience, while all the time she worked surreptitiously to ensure the women had that freedom. "The Superior General desires that if a confessor does not give comfort, he should be changed. But do not speak of this matter with any of the priests," she cautioned. Rather, "let it alone, that with time it may work itself out, and little by little the thing shall be done."[65] Recusant Mary Ward states yet another reason to allow women to choose their own confessor-directors: communication. The English nuns living in exile should have a confessor "who could speak their own language and understand and feel with their difficulties, especially at a time of great national distress."[66]

Jane Frances de Chantal makes her opinions about a confessor's domain quite clear. She has utmost respect for confessors, but insists that there is an explicit line demarcating women's dependence on confessors and their freedom from them. "We must show great respect to the confessors and do all we can reasonably to satisfy them, honoring God in them. However," she firmly clarified, "we must not be subject to them in such matters as procuring preachers, having Masses said, receiving holy communion from persons of repute or others whom we may want to please, or to confess ourselves to such persons when it seems suitable." Here she became quite insistent, for she considered the issue to be at the heart of their spiritual lives: "In all such matters you must remain very free, for such decisions rest completely with you. This is what the rule and our customs specify. And just as we must, with prudence and discernment, make use of the holy liberty that is given us, so must we guard it carefully and jealously, but always with humility. We show these confessors due respect, yet we explain to them very frankly our liberty of action."[67]

Finally, and perhaps most important, with the advent of the confessor-director women gained something still very much needed in this world of religious turmoil: protection from and access to male power structures. By placing themselves within the penitential system, women theoretically were subjected to their confessors, even though we have seen that, in fact, the degree of spiritual experience determined who was dominant in the director-directee relationship. The theoretically subjection, however, served the women well, for it created a buffer between them and their critics in society. Teresa of Avila wrote quite openly and appreciatively about the function of confessor-directors. She tells us of the "very great trials" that befell her confessor-director Baltasar Alvarez "on my account, and this in many ways. I knew they used to tell him that he must be on his guard against me" because of her advanced spirituality and reform work.

> For three years and more during which he was my confessor, I gave him a great deal of trouble with these trials of mine, for during the grievous persecutions which I suffered and on the many occasions when the Lord allowed me to be harshly judged, often undeservedly, all kinds of tales about me were brought to him and he would be blamed on my account when he was in no way blameworthy. Had he not been a man of such sanctity, and had not the Lord given him courage, he could not possibly have endured so much, for he had to deal with people who did not believe him but thought I was going to destruction, and at the same time he had to soothe me and deliver me from the fears which were oppressing me.[68]

After Angela Merici decided to engage in her pioneering educational apostolate among the diseased and unschooled poor, she waited until her confessor-director agreed with her decision to formally establish a religious order; only then did she proceed to seek approval from the official ecclesiastical hierarchy.[69] So insistent was she that the women have the protection of obedience to a confessor before engaging in new areas of ministry that it was encased in her *Rule:* "According to the judgement of their confessor, all of them must carry on some charitable work, especially that of Christian instruction."[70] Margherita Maria Alocoque developed a deep love of the Sacred Heart, but she first had to persuade her rather recalcitrant confessor, Claude de la Colombière, that her devotion was valid before she was able to spread the cult of the Sacred Heart and to initiate a feast for its celebration.[71] When Alix le Clerc became "filled with the spirit, which must be believed, to found a new order of girls to do all the good that can be done," she decided "to be like St. Ignatius, who inspired me to educate young girls, who in many cases were like delicate straw." In order to fulfill her goal, nevertheless, she had to have the church's approval, and the church's approval could be attained more easily if she persuaded her confessor to be her advocate. "This vocation was at first denied" because Pierre Fournier, her confessor, "did not understand this desire to 'gather together the poor delicate straw.'" Once she convinced him of the genuineness of her vocation and that "I would never abandon my intention,"[72] Fournier became such an active advocate for Alix that most historians traditionally have identified him as the founder of Alix's Canonesses Regular of St. Augustine of the Congregation of Our Lady.[73] Lack of historical credit was a price women apparently were willing to pay as long as they were able to achieve their own personal goals.

Innumerable women gained access into the literary world through their confessors and then had their writings receive protection from hierarchical scrutiny. Introductory, qualifying statements like María Vela's "I am writing this because you have so ordered under express obedience"[74] deflected ultimate responsibility from the woman to the confessor-director. It became the latter's job to see that the writings were approved and then circulated among the people. In the case of María Vela, her confessor-director, Francisco de Salcedo, took his job very seriously. He took it upon himself to journey throughout the country, seeking approval of María's writings. José de Acosta, rector of the Jesuit college in Salamanca; Luis de la Puente, retired teacher of philosophy and theology in Valladolid; an unnamed confessor at Santa Ana; Juan de Alarcón, lector of scripture at the Royal Convent of Santo Tomás; Pedro Martínez, professor of theology at Santo Tomás: All these men were asked for and gave their approval.[75] In this way María Vela was protected from criticism in ways she never would have been able to receive herself. Teresa even used obedience to her confessor to justify and protect her choice of subject matter: "As I have been commanded and given full liberty to write about my way of prayer and the favours which the Lord has granted me, I wish I had also been allowed to describe clearly and in full detail my grave sins and wicked life."[76]

Some women apparently were even taught to write as part of their spiritual direction. For example, Angela Mellini testified that her confessor-director, Giovanni Battista Ruggieri, asked her "in the confessional if I knew how to write; and after I told him no, he told me that I must learn how and he gave me a sample of letters on a page and a pen." He desired this "because he said that if he left Bologna he wanted to learn my spiritual progress from me."[77] Mary Ward doubted whether she had enough literary skill to write because she had "difficulty in finding fit words for what I would express," so she procrastinated "three or four years" before her confessor, Roger Lee, gave her on his deathbed "a more absolute charge to do it." He was so anxious to have her writing circulate that he attempted to protect it from the grave by placing her under obedience to "leave it sealed up with our company here whenever I was to undertake any fresh journey" such as to England "or any other place where my life or liberty might be endangered."[78]

As many scholars argue, however, most women used their confessors' commands to write to circumvent hierarchical criticism. "Often the impulse to write came first from the nun herself or from one of her Sisters," historians Electa Arenal and Stacey Schlau state. "Reported to the confessor, the impulse could then be transformed into a request from him—which the nun could meet with humility."[79] We see how the nuns of Port-Royal so arranged to have their superior, Angélique Arnauld, write her memoirs. They are completely open and even proud of their machinations.

> We had begun without her knowledge, more than two years previously, to write memoirs of all we had been able to learn relating to the early nuns of Port-Royal and to herself, touching what she had done to establish reform there and afterwards. But although we endeavored to get her to write about them. . . . she thrust us away when we tried to speak of her writing in good earnest. Since we saw accordingly that we gained nothing by persuasion we thought well to obedience to bear, for she never resisted that. We had recourse for that to M. Singlin (her confessor) and we implored him to ordain it on her. He did so.[80]

Given all these qualifications and limitations, the obvious question to ask is why confessor-directors not only tolerated the situation but also helped create it and sustain it. Fortunately, the majority of Roman clergy learned the hard lessons of the Reformation. If the Roman church wished to avoid another internal conflagration, it had to address the needs of its people as they presented themselves and before pressure built to dangerous levels. Encouraging the placement of the spiritual director within the penitential system, but not relegating the director exclusively to the system, provided society with a place where males and females could redefine their relationship to society and to each other in the changed world of post-Reformation Europe. In this world the Roman church was suddenly on the defensive, and it needed women as well as men to deflect the assaults of Protestantism and stem the tide of defections. The Jesuits have long been recognized as the chief agents of the Tridentine church in its effort to combat Protestantism, but not as well recognized is the role women played in this effort. The Council of Trent (1545–1563) realized that education of its clergy was essential if the church wished to survive, hence it mandated the establishment of seminaries for the training of its priests. At the local level the devout came to the same conclusion: Education of the laity in the truths of Catholicism was the best defense against Protestantism. Women were in the forefront of this movement. Soon schools for both boys and girls were springing up throughout the West. In Italy, even before the Council, Angela Merici established the Company of St. Ursula whose main purpose was to "carry on some charitable work, especially the work of Christian instruction."[81] In France, women dévotes energetically promoted the idea of the necessity of an informed laity, and to that end about eighteen new teaching congregations were established there by 1640.[82] In the Low Countries, English recusants, like Mary Ward's group, founded boarding schools "not only of our nation, of which there were a great many, but also those of the place where they lived [St. Omer], who were taught *gratis*, that all become good Christians and worthy women."[83] Eventually Mary Ward molded her group into the Institute of the Blessed Virgin Mary, and her first plan for the group articulated the reasons for women's involvement in the teaching apostolate.

> Since the very distressed condition of England, our native land, is greatly in need of spiritual workes, just as priests, both religious and secular, exercise an increasing apostolate in this harvest, so it seems right that, according to their condition, women also should and can provide something more than ordinary in the face of the common need.
>
> And since many women outside of England serve God most devoutly in monastic communities, and day and night by their prayers to God and good works, contribute very much towards the conversion of the kingdom, so we also feel that God (as we trust) is inspiring us with the pious desire that we should embrace the religious life and yet that we should strive according to our littleness to render to the neighbor the services of Christian charity, which cannot be discharged in the monastic life.
>
> Accordingly, we have in mind the mixed life, such a life as we learn Christ our Lord and Master taught his chosen ones, such a life as his Blessed Mother lived and handed down to those of later times, in those times especially when the Church was afflicted, as in our country now, in the Christian virtues and liberal arts so that they may be able thereafter to undertake more fruitfully the secular and monastic life, according to the vocation of each.[84]

That the clergy needed the women to minister to the people is key to understanding why such flexibility was allowed in the relationship between directors and directees. Women got what they wanted, and the clergy got what they wanted. It was, in short, a satisfactory, reciprocal relationship. The nature of the arrangement is clearly seen in Louise de Marillac's petition to the rector of Notre-Dame de Paris for permission to establish schools. In a formal application, Louise "very humbly supplicates" the rector, "informing him that the sight of the great number of poor in the Saint-Denis district leads her to desire to take charge of their instruction." The Roman hierarchy shared her concern. "Should these poor little girls remain steeped in ignorance, it is to be feared that this same ignorance will be harmful to them and render them incapable of cooperating with the grace of God for their salvation."[85] The rector's response reveals why the hierarchy granted women these new positions of authority. By allowing them this authority, the clergy solved a pressing problem without abdicating their own position of ultimate power. After being assured by "trustworthy persons who have knowledge of your life" that Louise was "worthy to operate schools," the rector gave her permission "on the condition that you teach poor girls only and do not accept others; that you educate them in good morals, grammar and other pious and honest subjects. You shall do all this after first swearing that you will faithfully and diligently operate these schools in keeping with our statutes and decrees."[86]

There were many other reasons why individual confessor-directors personally encouraged the new director-directee relationship and allowed their directees such leeway. Historian Jodi Bilinkoff discusses many of these reasons in her analysis of relationships between five early modern women from Avila—María de Santo Domingo, Mari Díaz, Teresa of Avila, María Vela, and Isabel de Jésus—and their confessors. "Male confessors were strongly attracted to the idea of directing spiritually advanced women, and, in turn, became deeply influenced by them, identified with them, and even became dependent upon them," because the relationship was an occasion for their own growth, spiritually, socially, and psychologically. Many underwent radical personal transformation as a result of their association with holy women, learning much from their intense spiritual experiences and gaining much from their revered social status. Bilinkoff also points out the advantages confessors saw in writing these women's vitae. For less sincere directors, it afforded them an opportunity to permanently link their names with saints, to expand their roles in the spiritual progress of their directees, to proclaim their expertise as directors and teachers, and to enhance their positions in the ecclesiastical world.[87] There were apparently enough disingenuous confessors so directing to merit a condemnatory remark from the esteemed Luis de la Puente, who cautioned those who attempted to "make a name for themselves through their penitents."[88] The more humble directors used their association with holy directees to further their own understanding of spirituality, to engage in theological disputes and controversial matters, and to learn from masters the intricacies of mysticism.[89] Some clergy actually owed the security of their vocation to these holy women.[90]

A glance at Teresa of Avila's writings finds numerous proofs of the latter thesis. Teresa tells us of a vision she had in which her confessor was spiritually rewarded for the aid he gave her work. "I saw Our Lady putting a pure white cope on a Presentado of this same Order [Pedro Ibáñez] of whom I have several times spoken. She told me that she was giving him that vestment because of the service he had rendered her in helping in the foundation of this house, and as a sign that from that time forward his soul would remain pure."[91] In another vision, Teresa saw a dove rest over a Dominican priest, which she took "to mean that he was to draw souls to God,"[92] and she "had a number of visions of the great favours which the Lord was bestowing" on the Jesuit rector Gaspar de Salazar, which "brought him great comfort and gave him courage" when Teresa told him about it.[93] In a letter to a mutual friend, Teresa took credit for one of her confessor's appointment to the chair of theology at Durando: "What do you think of the creditable manner in which Fray Domingo Báñez obtained the chair? God protect him, for I barely succeeded in winning it for him."[94]

Obviously, not all confessee-directees were holy women, and thus most confessor-directors could not anticipate reaping such benefits. Confessor-directors had power and control over many aspects of these women's lives, whether they actively sought and enjoyed it or passively accepted and were humbled by it. Historian Asuncion Lavrin has documented many such incidences where confessor-directors strongly influenced the vocational choices of women in seventeenth-century New Spain[95]; most such examples, however, would be unrecorded, given the likelihood that forced vocations would not blossom into enough spiritual or historical significance to be written about. Women who after attaining some spiritual status later recorded their earlier dealings with confessor-directors give us an indication of how many clergy were concerned with total control. We have already heard Ana de San Bartolomé's account of the battle between herself and Cardinal Bérulle over control of the Carmelite reform monasteries in France, and we have listened to Alix le Clerc's complains about Pierre Fournier's hesitancy to allot her enough power to form a new religious community.

Mary Ward had a long history of disagreements with confessor-directors concerning her actions, but in her narration of these conflicts, we plainly see that she was not willing to assign negative motives to her confessor-directors' desires to control her decisions. When her first confessor-director, Richard Holtby, "to whom I had confessed for seven years, was also of the opinion," along with her father, "that in no way ought I to leave England not make myself a religious," Mary ascribed good intentions to his directives, even though he used his authority to forbid her to fulfill what she knew was her vocation. "His words were of weight, and on this occasion caused me inexpressible distress, because I did not dare to do what he prohibited as unlawful, nor could I embrace that which he proposed as my greater good. His motives were pious, prudent and regardful of the service of God and the common good."[96] When she told another confessor-director, Roger Lee, about her decision to leave one religious community to found another, he, too, "appeared to disapprove, and oppose with a certain severity unusual to him. He exhorted me to a more than ever exact observance of the rules and regulations of the place where I was."[97] While ultimately Mary Ward's decisions were correct and bore fruit, we cannot automatically conclude that the confessor-directors' decisions were therefore wrong or attribute to the clergy a desire to control for control's sake. They may have even been helpful to Mary's long-term spiritual development. Indeed, they may have prepared her for the resistance her new order met on the international scale in the latter part of her life.

Protestant opponents of the Roman church often made much of this issue of control, particularly as it applied to Jesuits. A thorough analysis of Jesuit spiritual direction of women is yet to be done, but there is no doubt that many nineteenth-century Protestant apologists and historians centered their hatred of the Jesuits on their spiritual direction of women in the confessional. "The spiritual direction and government of woman is the vital part of ecclesiastical power, which the priest will defend even to the death. Strike, if you please, at any thing else, but not at that!" wrote Jules Michelet in his inflammatory *The Romish Confessional or the Auricular Confession and Spiritual Direction of the Romish Church, its History, Consequences and policy of the Jesuits.*[98] Michelet maintains that traditional spiritual "consolations were withdraw [*sic*] to the sixteenth century" from female monasteries, and spiritual directors were substituted, "something new, and little known in the middle ages, when they had only the confessor." The director controlled the nuns: "God, whom they had sought in their reading, and by their sighs was henceforth to be daily dispensed to them by this man." Because this control was so enticing, "during more than a quarter of a century, priests, and monks, the religious of all robes, had, among themselves, an active war" over who would "exercise undivided dominion" over the women.[99] Ignatius Loyola, however, had forbidden members of his order to take on the direction of women's monasteries.[100] Not to be undone by other clergy, the Jesuits, according to Michelet, found a way around Loyola's injunction: "The Jesuits did not govern the institutions collectively, they directed the sisters one by one."[101]

It was not just Jesuit control of religious women that Michelet denounced; he was more vehement in his denouncement of directors of married women. A director "now knows of this woman, what the husband has never known," and everyone "the master of the thoughts, is he to whom the person belongs." As a result, "an entire division is made between the spouses, for now there are two; the one has the soul, the other has the body." Indeed, this is the reason why the priest strives for control—"to weaken the bonds of family ties; to undermine a vital authority; I mean that of a husband."[102] The notorious anti-Catholic Pastor Chiniquy echoed Michelet's thesis: "In the Church of Rome, through the confessional, the priest is much more the husband of the wife than the man to whom she was wedded at the foot of the altar. The priest has the best part of the wife. He has the marrow, when the husband has the bones." Such a situation creates havoc in family life, for "will she not naturally, instinctively serve, love, respect, and obey, as lord and master, the godly man, whose yoke is light, so holy, so divine, rather than the carnal man, whose human imperfections are to her a sound of daily trial and suffering?"[103] Chiniquy then imposes Michelet's thesis on the whole of national histories. "Why is it that the Irish Roman Catholic people are so irreparably degraded and clothed in rags? Why is it that that people, whom God has endowed with so many noble qualities, seem to be so deprived of intelligence and self-respect that they glory in their own shame?" The answer is simple: "The principal cause is the enslaving of the Irish women, by means of the confessional. Every one knows that the spiritual slavery and degradation of the Irish woman has no bounds. After she has been enslaved and degraded, she, in turn, has enslaved and degraded her husband and her sons."[104] Likewise, "do not look for the causes of the downfall, humiliation, and untold miseries of France anywhere else than in the confessional." In short, "the nations who allow the women to be degraded and enslaved at the feet of her confessor—France, Spain, Romish Ireland, Mexico, etc., etc.— are, there, fallen into the dust, bleeding, struggling, powerless, like the sparrow whose entrails are devoured by the vulture."[105]

Before we dismiss such propagandists' claims as useless to the historian, we should contemplate Rene Fülöp-Muller's thesis that we "will find more valuable help in these partisan controversial writings than in the guaranteed information of the historian," because "all the hate filled pamphlets, the highly coloured apologies, distorted misrepresentations, doctored reports, the slanderings and glorifications" actually may do more to disclose how deeply something "influenced emotions, thoughts and actions at every period."[106] However exaggerated and misguided their analyses are, both Michelet and Chiniquy were correct in their identification of a new element present in the post-Reformation world. Early in the nineteenth century *The Secret Instructions of the Jesuits* was widely published and circulated throughout Europe[107]; supposedly it was an edition of a 1658 book discovered in the Jesuit archives in Westphalia. Here we find the same preoccupations as Michelet and Chiniquy fixate on, only this time disguised as a primary source proving the propagandists' thesis. The sole goal of the *Secret Instructions* was to provide Jesuits with rules as to how "to gain power; the acquisition of which has ever been the first object with this Society."[108] To this end "the confessor must manage his matters so, that the widow may have such faith in him, as not to do the least thing without his advice, and his only."[109] In fact, "let this be deeply imprinted on their minds, that, if they desire to enjoy perfect peace of conscience, they must, as well in matters temporal as spiritual, without the least murmuring, or inward reluctance, entirely follow the direction of their confessor."[110] Women other than widows will come under the control of the Jesuit directors if they "let women that are young, and descended from rich and noble parents, be placed with those widows, that they may, by degrees, become subject to our directions."[111]

What we can deduce from much of this is that many segments of society were keenly aware of the potential the new figure of the confessor-director possessed. Women emphasized mostly the positive aspects for themselves as directees and even as directors. Beyond their personal involvement, women generally saw the emphasis on spiritual direction to be positive. Protestant propagandists saw the confessor-director as inherently dangerous because of the power the figure had to control one's inner life. That it gave a woman access to a man other than her husband and inversely that a man other than a husband had access to a wife often was seen to be radically dangerous to patriarchal control. Roman clergy were aware that the confessor-director provided a way to advance themselves temporally or spiritually or both. With the historical distance we now have, we see that each group's judgment, at different times and in different societies, contained to varying degrees at least some element of truth. There are two further conclusions, though, that must be acknowledged. Regardless of how complex and confusing the advent of confessor-spiritual directors was, in no way did it hinder or preclude the presence of women spiritual directors. They continued their traditional ministry without interruption. Second, we find society gradually realizing the validity of one of this study's basic theses: Those that articulated the meaning of life and directed others to it were indeed powerful historical forces.

Chapter 7

$\diagup\!\!\!\!\!\diagdown$

Spiritual Directors
within the New Orders

In the first chapter we discussed the groundbreaking role that the Samaritan woman played in the ministry of Jesus; immediately after comprehending the meaning of Jesus' message, she departed to direct others to their salvation. As a result, "many Samaritans from the town believed in him of the strength of the woman's testimony" (Jn 4:39). Lest the skeptic think that such exegesis is the product of overzealous projections into Scripture on my part, we should listen attentively to Teresa of Avila's ruminations on the Samaritan woman's spiritual direction.

> I have just remembered some thoughts which I have often had about that holy woman of Samaria, who must have been affected in this way. So well had she understood the words of the Lord in her heart that she left the Lord Himself so that she might profit and benefit the people of her village. This is an excellent example of what I am saying. As a reward for this great charity of hers she earned the credence of her neighbors and was able to witness the great good which Our Lord did in that village. This, I think, must be one of the greatest comforts on earth—I mean, to see good coming to souls through one's own agency. . . . To me the astonishing thing is that they should have believed a woman—and she cannot have been a woman of much consequence, as she was going to fetch water. . . . In the end, her word was believed; and merely on account of what she had said, great crowds flocked from the city to the Lord.[1]

Teresa quite clearly was aware of the scriptural foundation for women spiritual directors. She emphatically argued that the Samaritan woman's ability to guide people was due not to social or financial status but purely because of the truth of her message. The significance of the message in fact made all aspects of the messenger insignificant, including her sex. It is also clear that Teresa appreciated how rewarding it is to contribute to another's spiritual advancement. "What a great thing it is to understand a soul!" Teresa admitted humbly.[2] Such skill was a gift, and if one possessed the necessary qualities spiritual directing was both obligatory and self-satisfying. When the nuns at St. Joseph's Monastery learned that Teresa had leave to write about prayer, their desire to obtain her guidance was so great that they "so earnestly begged me to say something to them about this that I resolved to obey them," Teresa records. Because they freely desired her spiritual advice, Teresa believed that "what I shall say to them" would be "more acceptable than other books which are very ably written."[3]

Teresa apparently thought her being a woman was more, not less, of a reason of offer direction to nuns. In a comment typical of what Alison Weber has called her "rhetoric of femininity,"[4] Teresa wrote in her preface to *Way of Perfection* about her suitability to direct nuns: "I know that I am lacking neither in love nor in desire to do all I can to help the souls of my sisters to make great progress in the service of the Lord. It may be that this love, together with my years and the experience which I have of a number of convents, will make me more successful in writing about small matters than learned men can be. For these, being themselves strong and having other and more important occupations, do not always pay such heed to things which in themselves seem of no importance but which may do great harm to persons as weak as we women are."[5]

It was the realization that she had such insight into life's mysteries that made Teresa convinced she must share it with others. For example, when she found the key to understanding the profundities of the Song of Solomon, she felt obligated to write it down for the benefit of others. "For about two years past, it seemed to me that for a purpose of my own the Lord has been enabling me to understand something of the meaning of a few of these texts, which I think will bring comfort to those sisters whom the Lord is leading along this road, as well as to myself."[6] Thus Teresa directed, regardless of the potential risks involved in glossing the difficult Song of Solomon (Luis de Léon was arrested in 1572 in part because of his translation of the book into the vernacular). The commentary, known of *Conceptions of the Love of God,* was the only one of Teresa's writings known to have been burned by one of her confessors because of the risk involved,[7] specifically, because he thought "it a new and dangerous thing that a woman should write on the Songs," according to Gracian in his preface to the first published edition of *Conceptions.*[8] Anticipating such reaction, Teresa argued in the first chapter of *Conceptions* that gender was not sufficient enough reason to outlaw her from writing a scriptural commentary. "Just so, we women need not entirely refrain from enjoyment of the Lord's riches" to be found in Scripture, Teresa proclaimed, because God can and does talk to and through women. In the past God "has been pleased to allow me to succeed in explaining other things to you (or perhaps it was His Majesty Who explained them through me, as they were intended for you)," so Teresa believed it proper for her to explain Scripture to others. While she emphatically stated that "I shall interpret these in my own way," she also insisted that "it is not my intention to suggest that I shall discover the truth" about the Songs apart from "what the Lord has to teach me." She willingly admitted that her commentary was valid only as long as "I do not depart from what is held by the Church and the Saints."[9]

In *Way of Perfection* (the second part of which, although overlooked by most critics, is a commentary on the Paternoster, a scriptural prayer), she articulates one of her most earnest and persuasive pleas for women's involvement in spiritual ministries such as direction of souls. "Hear us not when we ask Thee for honours, endowments, money, or anything that has to do with the world; but why shouldst Thou not hear us, Eternal Father, when we ask only for the honour of Thy Son," Teresa implored. She had total confidence in her and in all women's role in ministering to people, first, because "when Thou wert in the world, Lord, Thou didst not despise women," but "didst find more faith and no less love in them than in men," and, second, because "it is not right to repel spirits which are virtuous and brave, even though they be spirits of women."[10] In a final but revealing comment Teresa reinforced how satisfying spiritual direction is: "Although I may be saying a lot of stupid things, I like telling you about my meditations, as you are my daughters."[11]

Although some of Teresa's contemporaries may have needed her rhetorical arguments to persuade them to consider her spiritual direction without prejudice, once her direction was read and circulated its validity was apparent to all. It is nigh impossible to overemphasize the contributions Teresa made in the realm of spiritual direction, both directly upon her directees and indirectly through the influence her writings had on all future spiritual direction. Indeed, between Teresa's analysis of the spiritual life and Ignatius Loyola's methodology for developing a spiritual life,[12] spiritual direction became firmly established within the Roman church. Emphasizing Teresa's magisterial contributions to spiritual direction does not necessarily diminish the roles of those like Francis de Sales, traditionally designated as the initiator of a golden age of spiritual direction, Cardinal Bérulle and the French School, Ignatius Loyola and his Society, or John of the Cross, the mystic's mystic and poet laureate of the spiritual life.

Giving Teresa her proper place in the history of spiritual direction helps us focus our attention on two crucial facts. With the exception of Loyola, first, chronologically Teresa precedes all the great masters, and, second, they all acknowledge their debt to her. Francis de Sales cited Teresa frequently, used her threefold classification of prayer in his spiritual direction, and was repeatedly exposed to Teresan spirituality by French advocates such as Madame Acarie and Jane Frances de Chantal. He directed others to read Teresa's works and follow her example as he did himself, and, significantly, Francis cited Teresa in his masterpiece, *Introduction to a Devout Life*, and in his letters as his authority for the definition of the proper director-directee relationship.[13] It is particularly evident that Francis formed his own spirituality in an environment already saturated with Teresan ideals. He acknowledged to Marie Brûlart that he was "delighted" about her adherence to "Mother Teresa, whom you love so much."[14] The cofounder of what we today call Salesian spirituality was Jane Frances, and she first learned of the prayer of quiet from the Teresan reform Carmelites at their Dijon monastery, which Jane frequented and contemplated joining early in her spiritual quest.[15] Toward the end of her life Jane still considered the Carmelites at Dijon her model, urging her novices to "let us do the same."[16] Teresa's *Way of Perfection*, a book Jane "highly approve[d] of," was also the only book Jane specifically asked novice mistresses to read to their charges, "for it is very useful, and may well help and incite them to the love of these two virtues, mortification and prayer."[17] Likewise, Cardinal Bérulle achieved his reputation in spiritual circles to a large extent because of his association with Madame Acarie's circle of dévotes, all of whom were highly influenced by Teresa of Avila. Teresa's works were translated and published in France toward the end of the sixteenth century, and, in historian Elizabeth Rapley's words, "the effect was electric."[18] Bérulle was one of the earliest clerical champions of Carmelite women, and by the mid-seventeenth century there were fifty-six Teresan reform monasteries throughout France. By the end of his life Bérulle's spirituality was distinctly at odds with Teresan spirituality, but there is no denying that his own interpretations and development were enhanced by both what he accepted in Teresan spirituality and in what he rejected.[19]

John of the Cross first encountered Teresa and her spiritual direction at the very beginning of her career as a confessor-director, only months after his ordination in 1567. Within a year of their first meeting Teresa found a farmhouse in Duruelo for John to establish the first reform monastery for Carmelite friars. To learn Teresan spirituality more thoroughly, John apprenticed himself to Teresa in a new foundation in Valladolid, where he remained until she was satisfied with his training.[20] Having thus learned from the master, John's own spiritual direction developed from this firm foundation. That John, like Teresa, devoted his energies to the detailed analysis of the life of prayer says more about his debt to Teresa than perhaps any of the many similarities in their spiritualities.

Teresa's contributions to spiritual direction were in many ways revolutionary. This is not due to any new methodology or to a redefinition of the relationship between director and directee, but rather because it was compatible to the revolutionary changes of the era. The Teresan spiritual director must identify, dissect, and analyze the elements of the spiritual life, just as the early modern intellectual was identifying, dissecting, and analyzing the physical elements of all created life, human and nonhuman. As the heavens yielded its secrets to the likes of Nicholas Copernicus and Galilei Galileo, and nature yielded its mysteries to William Harvey and Robert Boyle, so, too, the intricacies of the spiritual life yielded its secrets to Teresa. She offered society what it craved, a science of the interior life.[21] The conflicts of the reformers and counter-reformers had perhaps rendered Christian theology sounder and more comprehensible to contemporaries, but, unfortunately, the same was not true for spirituality. Traditional pietistic devotions were questioned and sometimes forbidden by reformers, and counter-reformers themselves were often at a loss as to how to discern the good from the bad when it came to many manifestations of spirituality. This was an era of witches, false prophets, confused messages, hate-filled propaganda, and spiritual adventurers. How could one be properly directed to the ultimate goal in life if one could not even be sure the director understood what the ultimate goal was? Teresa gave spiritual direction new life by first giving spirituality new life. She did this by scrutinizing her own experience and understanding of life and writing clearly about it in a language society comprehended. John of the Cross's analysis of the dark night of the soul may have been more profound and Francis de Sales's approach more accessible to the laity, but it was Teresa who set spirituality and spiritual direction on the proper course for the new era.

It is not by chance that Teresa constructed a spirituality and spiritual direction suited to the needs of her society; rather it was the result of careful reflection upon her contemporary world. How conscious and sensitive she was to these issues is particularly evident in her opening chapters of *Way of Perfection*. After acknowledging the disorder theological conflicts had wreaked upon nearly every aspect of life, Teresa addressed the question of proper responses as a paradigm. She tells us that at about the same time she made her decision to found her first reform convent, "there came to my notice the harm and havoc that were being wrought in France by these Lutherans." While "this troubled me very much," she was especially frustrated because as "a woman, and a sinner" she was "incapable of doing all I should like in the Lord's service."[22] As we read on, however, we find that the limitations placed on her as a sinful woman are not the real issue. She is, instead, questioning the social utility of the contemplation. In the face of endless work that must be done in the world, how can the contemplative, enclosed nun justify her apparent failure to pursue solutions actively? The reformers' attack on monastic life—a life that embodied intense concentrated attention to the interior life of a person—demanded an answer if it were to remain viable. Teresa responded to the criticism by identifying the specific role the contemplative plays in a world increasingly filled with spiritual activism: "I determined to do the little that was in me—namely, to follow the evangelical counsels as perfectly as I could, and to see that these few nuns who are here should do the same" and thereby "be able to give the Lord some pleasure, and all of us, by busying ourselves in prayer, for those who are defenders of the Church, and for the preachers and learned men who defend her, should do everything we could to aid this Lord of mine."[23] Society's criticism of the monastic life should not surprise the nuns, for "since worldly people have so little respect for Thee, what can we expect them to have for us?" Even though the worldly "have won severe punishment at his hands," Teresa confessed that "it breaks my heart to see so many souls travelling to perdition." Because she and her fellow nuns were dedicated to the pursuit of perfection and because this pursuit gave "the Lord some pleasure," they must use their position to save all those "being lost every day" to worldly, transitory, and ultimately meaningless goals. "Oh, my sisters in Christ!" Teresa implored. "Help me to entreat this of the Lord, Who has brought you together here for that very purpose. This is your vocation; this must be your business; these must be your desires; these your tears; these your petitions. Let us not pray for worldly things, my sisters," Teresa concluded, for "the world is on fire" with people erroneously concerned with "things of little importance."[24] The contemplative must concentrate solely on things of eternal importance.

Teresa is keenly aware of the increased complexities of society, of the lack of consensus concerning life's priorities, and of the need for diverse vocational responses to these realities. She insisted that one found meaning in one's life by fulfilling the role one is best suited for. In order to complete her specific role, the nun must place her physical well-being in the hands of God and "keep your eyes fixed upon your Spouse: it is for Him to sustain you." The nun must not worry about income but trust simply and completely. That is not to say everyone must do likewise to attain happiness: "Let those whom the Lord wishes to live on an income do so: if that is their vocation, they are perfectly justified; but for us to do so, sisters, would be inconsistent."[25] A nun's freedom from worldly and physical cares, though, is a necessary luxury so her primary task can be successfully completed. Teresa repeats her understanding of the task, this time using a metaphor. "The present day," she begins, "is like a war in which the enemy has overrun the whole country, and the Lord of the country, hard pressed, retires into a city, which he causes to be well fortified, and whence from time to time he is able to attack."[26] It is the contemplative nun's job to pray to God "to make the captains in this castle or city—that is, the preachers and theologians—highly proficient in the way of the Lord" and to "pray that they may advance in perfection, and in the fulfillment of their vocation, for this is very needful. For, as I have already said, it is the ecclesiastical and not the secular arm which must defend us. And as we can do nothing by either of these means to help our King, let us strive to live in such a way that our prayers may be of avail to help these servants of God, who, at the cost of so much toil, have fortified themselves with learning and virtuous living and have laboured to help the Lord."[27]

Thus, for Teresa life must be lived with both individual and collective responsibilities in mind. In fact, only when one's responsibilities to the community are met is it possible to meet one's personal responsibilities; they are inseparable. "Some people think it is a hardship not to be praying all the time for their own souls," Teresa commiserates. "Yet what better prayer could there be"[28] than the prayer that fortifies the captains of the castle who "have to do business with the world, to live in the world, to engage in the affairs of the world, and, as I have said, to live as worldly men do, and yet inwardly to be strangers to the world, and enemies of the world, like persons who are in exile—to be, in short, not men but angels? Yet unless these persons act thus, they neither deserve to bear the title of captain nor to be allowed by the Lord to leave their cells."[29]

In this way the nun's individual prayers and sacrifices contribute to the well-being of all. "If we can prevail with God in the smallest degree about this, we shall be fighting His battle even while living a cloistered life and I shall consider as well spent all the trouble to which I have gone in founding [St. Joseph's Monastery]."[30] Eternal happiness comes through unselfishness. "You may be worried because you think it will do nothing to lessen your pains in Purgatory, but actually praying in this way will relieve you of some of them and anything else that is left—well, let it remain. After all, what does it matter if I am in Purgatory until the Day of Judgment provided a single soul should be saved through my prayer? And how much less does it matter if many souls profit by it and the Lord is honoured!"[31] The nun is not to be even momentarily bothered that her prayer is of less value because it comes from a woman, for God is "a righteous Judge, not like judges in the world, who, being, after all, men and sons of Adam, refuse to consider any woman's virtue as above suspicion"[32]; the nun is only to concern herself with the task she accepted. It is in her role as spiritual director that Teresa offered the concluding advice: "If your prayers and desires and disciplines and fasts are not performed for the intentions of which I have spoken, reflect (and believe) that you are not carrying out the work or fulfilling the object for which the Lord has brought you here."[33]

Much of what Teresa wrote about in her works is but guidance given to the contemplatives inside her metaphorical city under siege on how best to achieve their goal. As a good spiritual director Teresa was always keenly aware of the specific dimensions and historical context her directees were laboring under. In *Way of Perfection* she referred repeatedly to contemporary attacks on Roman interpretations of life's mysteries and the role nuns must play in defending that interpretation. "There must be someone, as I said at the beginning, who will speak for Thy Son, for He has never defended Himself. Let this be the task for us," Teresa cajoled her nuns.[34] Her reason for devoting so much of her attention to spiritual direction was intimately tied to her responsibilities to society at large. "By the earnest desire that I have to be of some use in helping you to serve this my God and Lord, I beg you, in my own name, whenever you read this," Teresa confessed in her conclusion of her spiritual treatise on prayer, *Interior Castles*, "to give great praise to His Majesty and beg Him to multiply His Church and to give light to the Lutherans."[35] She believed that knowledge was the key to successful completion, both self-knowledge and knowledge of the intricacies of the spiritual life that the women dedicated themselves to. "It is no small pity, and should cause us no little shame, that, through our own fault, we do not understand ourselves, or know who we are," Teresa lamented. "Would it not be a sign of great ignorance, my daughters, if a person were asked who he was, and could not say, and had no idea who his father or his mother was, or from what country he came? Though that is great stupidity, our own is incomparably greater if we make no attempt to discover what we are."[36]

Knowing we have a body and soul is not enough; we should also know "what good qualities there may be in our souls, or Who dwells within them, or how precious they are."[37] Such knowledge is, unfortunately, difficult but necessary. Indeed, "it is absurd to think that we can enter Heaven without first entering our own souls—without getting to know ourselves"[38]; therefore, "since this is so important, sisters, let us strive to get to know ourselves better and better, even in the very smallest matters."[39] Indeed, the consequences of not being introspective were dire. "There are souls so infirm and so accustomed to busying themselves with outside affairs that nothing can be done for them, and it seems as though they are incapable of entering within themselves at all," Teresa warned, and "unless they strive to realize their miserable condition and to remedy it, they will be turned into pillars of salt for not looking within themselves."[40] Those who fail to cultivate self-knowledge will never make progress toward happiness. "Self-knowledge is so important that, even if you were raised right up to the heavens, I should like you never to relax your cultivation of it," Teresa directed her nuns, for no matter "however high a state the soul may have attained, self-knowledge is incumbent upon it." According to Teresa, because the First Mansion of her *Interior Castle* begins with "the first rooms—that is, the rooms of self-knowledge,"[41] no further progress can be made without this introspection.

In fact, the halfhearted are saved from "great peril" only if they "know themselves well enough to realize that they are not going the right way to reach the castle door."[42]

While not as crucial as self-knowledge, knowledge of the interior life is most helpful to anyone aspiring to great happiness. At a fundamental level it clears up the frustrations humans experience in the face of ignorance. "I fully realize how important it is for you that I should explain certain interior matters to the best of my ability," Teresa states, "for we continually hear what a good thing prayer is, and our Constitutions oblige us to engage in it for so many hours daily, yet they tell us nothing beyond what we ourselves have to do and say very little about the work done by the Lord in the soul—I mean, supernatural work."[43] The nature of Teresa's contributions to society, however, reaches well beyond the soothing of certain frustrations; her writings are a clear, resounding clarion call to enter into a new and deeper level of human understanding of the pursuit of happiness. Yet for all their timelessness and universality, her writings also are grounded firmly in her own immediate world. Just as Teresa's emphasis on self-knowledge corresponds to early modern humanists' call for awareness of individuality, so, too, does her exploration of the unknown in the spiritual life correspond to society's exploration of uncharted territory. Here is a message responsive to the needs of her day, but her insight is so penetrating and true that it also is a message applicable to needs of people well beyond her own historical context. She wrote because "the nuns of these convents of Our Lady of Carmel need someone to solve their difficulties concerning prayer,"[44] but she also wrote because she believed that it was universally true that "if we turn from self towards God, our understanding and our will become nobler and readier to embrace all that is good: if we never rise above the clough of our own miseries we do ourselves a great disservice."[45] Proper development of prayer (a synonym for the interior life for Teresa[46]) was the way to rise above our frailties.

Anyone hoping "to walk along the way of prayer must of necessity practise" three things, according to Teresa in *Way of Perfection:* love of neighbor, detachment from created things, and humility, which "is the most important of the three and embraces all the rest."[47] There is nothing out of the ordinary in this list; love, detachment, and humility are traditional Christian virtues that many in the past have identified as necessary for personal happiness. Teresa's contribution lies in her awareness that these virtues bring salvation not only to the person but also to the world. Those involved in the business of the world—Teresa's "captains"—are not always free to immerse themselves in deep prayer. In Europe's new era of prosperity and activity, the way to integrate one's worldly life with one's inner life was not always evident and certainly not easy. Those who are given the opportunity to find the means for such integration must pray for themselves and for those who provide them with such an opportunity.

> You may ask why I emphasize this so much and why I say we must help people who are better than ourselves. I will tell you, for I am not sure if you properly understand as yet how much we owe to the Lord for bringing us to a place where we are so free from business matters, occasions of sin and the society of worldly people. This is a very great favour and one which is not granted to the persons of whom I have been speaking, nor is it fitting that it should be granted to them; it would be less so now, indeed, than at any other time, for it is they who must strengthen the weak and give courage to God's little ones. . . . Do you think, my daughters, that it is an easy matter to have to do business with the world? . . . Do not think, then, that they need but little Divine favour in this great battle upon which they have entered; on the contrary, they need a great deal.[48]

It is because the captains of the world are in such demand that Teresa's women must strive to offer the purest type of prayer possible, so "as to be worthy to obtain two things from God"; first, that there may be many leaders of the world "who have the qualifications for their task"; and, second, that they may work in the world without succumbing "to the song of the sirens."[49] In *Interior Castle* Teresa discussed the apostolic nature of prayer repeatedly.[50] She chastised those "people very diligently trying to discover what kind of prayer they are experiencing and so completely wrapt up in their prayers that they seen afraid to stir" and "think that the whole thing consists in this. But no, sisters, no; what the Lord desires is works."[51] Teresa deemed this point important enough to make it her thesis in the *Castle's* concluding chapter: "Believe me, Martha and Mary must work together when they offer the Lord lodging."[52] Indeed, Teresa maintains that during the prayer of quiet people "see clearly that their whole self is not in what they are doing." While their will "is united with its God," the remaining "faculties are left free to busy themselves with His service." Thus, the favor "unites the active life with the contemplative. At such times they serve the Lord in both these ways at once; the will, while in contemplation, is working without knowing how it does so; the other two faculties are serving Him as Martha did. Thus Martha and Mary work together."[53]

For those willing to do all that is necessary to practice love, detachment, and humility, the happiness of prayer awaits them, which in turn is the door through which one enters Teresa's most famous metaphor, *los morados,* the mansions.[54] In *Interior Castle* she describes seven mansions, each with innumerable rooms to accommodate the number of ways God can lead a person to happiness. Here Teresa offers her fullest analysis of the spiritual life, describing for her directees the intricacies of the prayer of recollection, of quiet, and of union, each stage being a preparation for the next. It is in the last stage that the person realizes the goal of humanity, happiness. "The soul is conscious of having reached this stage of prayer, which is a quiet, deep and peaceful happiness of the will, without being able to decide precisely what it is, although it can clearly see how it differs from the happiness of the world. To have dominion over the whole wold, with all its happiness, would not suffice to bring the soul such inward satisfaction as it knows in the depths of its will."[55] Unfortunately, because "few of us prepare ourselves for the Lord,"[56] few are chosen to experience the happiness of heaven, where that union of creator and creature "takes place in the deepest centre of the soul."[57] This earthly preview of ultimate happiness is "so sublime a favour, and such delight is felt by the soul, that I do not know with what to compare it, beyond saying that the Lord is pleased to manifest to the soul at that moment the glory that is in Heaven, in a sublimer manner than is possible through any vision or spiritual consolation."[58] All Teresa's spiritual direction, all her reforms, and all her activity has this one intention, to help one attain the happiness of perfection.

> God is happy, since He knows and loves and rejoices in Himself without the possibility of doing otherwise. He is not, nor can He be, free to forget Himself and to cease to love Himself, nor would it be perfection in Him were He to be so. Thou wilt not enter into thy rest, my soul, until thou becomest inwardly one with this Highest Good, knowing what He knows, loving what He loves, and enjoying what He enjoys. Then shalt Thou see the end of the mutability of thy will; then, then shall there be an end of mutability. For the grace of God will have wrought so much in thee that it will have made thee a partaker of His Divine nature, with such perfection that thou wilt neither desire nor be able to forget the Highest Good, nor cease to rejoice in Him and in His love.[59]

The greatest and most immediate impact Teresa's spiritual direction had was, of course, among her own nuns. According to María de San José, close confidante of Teresa's, all the women of the reform movement were indebted to Teresa's direction, because "the fruit springs forth in praise of the tree which produced it; and so anything I may say of the virtues and graces of the Sisters must be understood to have been achieved through [Teresa's] clear intellect and heroic virtue."[60] Teresa's influence, though, is not limited to the nuns, "for not only has she roused weak women to take up Christ's Cross, but she has shamed the men, and dragged them out to the field of battle; and when they had turned their backs on discipline and primitive virtue, made them follow the banner of their woman Captain, so that they may face their enemies who had risen to become so lordly. She began like another Deborah to inspire the army of God," and she did so

with unsurpassed genius: "For how could one describe how witty and tact-
ful she was, and how loving and gentle in manner; how prudent and wise,
with the caution and simplicity of a dove; how describe her faith and hope
and spirit of prophecy, the grace given her of bringing souls to God, her
marvelous gifts of counsel?—for, indeed, many of the nobles of Spain took
her advice in the gravest of matters."[61] Teresa's direction was so inspiring
that it was "what made me follow [the Carmelites]." María de San José
concluded "that if those whose work it is to bring souls to God were to use
the same schemes and skills used by the Saint, many more women would
come to religion."[62]

When reflecting further on Teresa's spiritual direction, María reveals how
much she imitated Teresa's approach to direction and endorsed her belief in
the obligation of nuns to direct others to God via their apostolate of prayer.
Like Teresa, María believed a rich spiritual life is build on a foundation of
self-knowledge; consequently María was often saddened by "how many
waste their time to no advantage because they do not know themselves" and
thus were incapable of "the heavenly practice of prayer."[63] Those who were
aware of how indispensable self-knowledge was to happiness must share
their awareness with those who were not, and also share their knowledge of
other means to achieve union with God. Such sharing should be second nature to the earnest, as María herself confesses to her directees: "And I tell you
Sister, that the best impulse I have had in this life has been to go through
the streets, to undeceive those who think it is a hardship to serve God."[64]
María continued thus: "I speak of you and me, and of all the rest of us whose
obligation it is to bring souls to God by our good example, and who pride
ourselves on being of Christ's own flock and of those who communicate
most closely with Him: we should not appear so gloomy and sorrowful, lest
we misrepresent our conversation with God to those who have not experienced it, giving them to understand that the practice of prayer, silence, and
spiritual exercises is a melancholy and unbearable thing."[65] Indeed, the nuns'
conversation with God takes place not only for their own sakes but also the
guidance of others toward the spiritual. María argued that the spiritual journey of "Saint Catherine of Siena, Saint Elizabeth, Saint Brigid, and the
Blessed Angela of Foligno and others" were told and recorded so everyone
should "know of all the delights and favors the Lord has bestowed on so
many sainted women." This is, in fact, the reason "why such favors are given,
and not solely to benefit those who receive them."[66]

María de San José summarized her interpretation of Teresan spiritual direction in a treatise for women spiritual directors of Carmelite novices, *On
the Instruction of Novices.*[67] Ana de San Bartolomé, another close associate of
Teresa's, wrote a series of works on spiritual direction[68] that also follow
Teresa's lead. Ana even tells us about two incidents in her life that reaffirmed

her faith in Teresan spiritual direction. One incident has already been discussed in the previous chapter, that which occurred when a confessor told Ana her impulse to aid souls was from the devil. When Ana checked with Teresa, however, she was assured that the confessor was incorrect and simply limited in his comprehension of the spiritual life. [69] Another assurance came after Teresa's death. When she told a confessor that "something close to my soul" was "an idea of Madre Teresa's," the priest told Ana "don't be like her." Upset that Teresa was not properly appreciated, Ana prayed "and the Lord came" to reassure her that Teresa's model was correct. The Lord "showed me the saint in glory, when he carried her under his arm, as if she had become one with him. And the Lord said to me: 'You see? Here I have her for you; none of this should bother you, let them say whatever they wish.'"[70] It was Ana de San Bartolomé who championed Teresan spirituality and direction in her own Carmelite order when division arose over adherence to Teresan rules and in France when confronted by Pierre de Bérulle.[71]

Teresa's influence in Spain quickly spilled over into neighboring France. Madame Acarie and her circle were heavily influenced by Teresa's writings on spirituality, recently translated and widely circulated among the Parisian group. An informal association of women who called themselves the Congregation of St. Genevieve met within the Acarie household, while male members of the group attempted to get formal approval to bring Teresa's reformed Carmelites to France and show the St. Genevieve Congregation how to found a Carmelite house. Francis de Sales went to Rome for papal permission, and, once granted, Pierre Bérulle, René Gauthier, and Quintanadoine de Brétigny proceeded to Spain to negotiate the transfer of nuns. Negotiations broke down temporarily when Bérulle insisted upon having nuns who were personally the spiritual directees of Teresa; agreement was reached when he received permission to bring six Carmelites, including the above-mentioned Ana de San Bartolomé and another of Teresa's closest directees, Ana de Jesus.[72]

Once in France, Teresan direction mixed freely with Ignatian methodology and native spiritual genius, and gave birth to such direction as that promoted by Jane Frances de Chantal and Francis de Sales. Together Jane and Francis created the popular Salesian spirituality that had as one of its chief characteristics an emphasis on accessible spiritual direction for laity and religious alike. Teresan influence is easy to find in Jane's spiritual direction, but Jane's own contributions are also abundant. Teresa's insistence that the practice of love, detachment, and humility is fundamental to all spiritual perfection is echoed time and again in Jane's *Instructions* given to novices. Love for the novice must begin with love toward the novice mistress. "Oh! you must love her much and have great confidence towards her and great sincerity in emptying your hearts into hers, thoroughly,"[73] Jane cajoled her novices, for a spiritual director can help one discover "all the means proper for arriving at the highest perfection."[74] Because "the spirit of the world and of the flesh cannot dwell with the spirit of religion and the spiritual life," the directees "must give up yourselves so completely into the hands of those who direct you, that they may twist you as they will, as we do with a handerchief."[75] Thus will the director help "destroy their old inclinations, habits, and propensities" that inhibit pure love and bring the directees along this path to living "an eternal life in Your Spouse."[76] One must remember, however, that "we can never attain to the perfection of holy love and union with God, unless we have also love of our neighbor."[77] In her sermons on the Rule of St. Augustine given to the Visitation chapter, she states this mandate unequivocally: "You see, my sisters, that in the Rule of St. Augustine proposes to us, in the first place, the great commandment of God, and tells us that *God is to be loved before all things*, and *after Him our neighbor.* The commandment then must needs be the foundation and the basis of our perfection; for in the observance of this lies the sum of Christian and religious perfection."[78] To Jane, then, love of one's spiritual director leads to love of one's neighbor, which in turn is essential to love of God and happiness in that union.

Detachment and humility follow the pursuit of pure love as naturally for Jane as they do for Teresa. Women enter religious life "to unite yourselves to God, and, to do this the better, to disunite yourselves from all that is not God." Jane's advice on how to accomplish this is clear: "Firstly, you must die to yourselves; that is, you must labor with courage and faithfulness at your perfection. Secondly, you must let others do their part, allowing yourselves to be flayed, stripped, and to have your heart bent."[79] In another Instruction where she used Jesus' scriptural mandate, "If any man will come after Me, let him deny himself," Jane further explains the extent detachment is necessary for pure love to grow. Detachment means one must "renounce all the wishes of the flesh, all our inclinations, desires, contentments, satisfactions, seekings of self, tastes, pleasures, fancies, habits, propensities, aversions, and repugnances, in things that are hard; in a word, to renounce in all and everywhere this evil self of ours. This renouncing, my dear Daughters, is the exercise which you are occupied in, and to which you must apply yourself during your novitiate, if you wish to begin to follow our Lord."[80] A person detached from self and so "stripped of the world, that strives to mortify the flesh, and is disinterested, will arrive, in a short time, at the very high perfection of loving union with its God, which is the treasure of treasures."[81] Likewise, a soul in such a detached, naked condition is humbled by true self-knowledge that "we are nothing, so that God may be our all; and the high point of humility is to desire, to accept, and to be glad to see oneself rebuffed" and "treated by all the world as worthless, abject, and as mere nothings." Rather than being dehumanizing, true humility "gives to the soul that possesses it true and holy liberty. It opens the heart to receive all from the hand of God."[82] The humble soul "holds itself so low and insignificant in the sight of God and of creatures" that union in love with God is made possible. Humility is for Jane as it was for Teresa and so many before her, "the mother of all" virtues. "Oh," she exclaimed, "if we could once acquire this virtue, we should soon have all virtues."[83]

Jane also promoted other aspects of Teresa's spiritual program. Self-knowledge, for example, was of key importance in the pursuit of happiness, according to Jane. "In truth, my dear Daughters, it is for want of knowing ourselves that we are astonished to see that we fall short," she reminds her nuns, "and Our Lord even permits us sometimes to have a very heavy fall, in order that we may become acquainted with ourselves." All progress toward perfection must commence with "knowledge of self" because it "will make us see many things in us to correct and reform." This knowledge "consists not in sentiment, nor in making great considerations upon it, but in believing it as being a truth of faith" that "we are weak, infirm, frail and far from perfect." Here Jane specifically called upon novice mistresses to train all future Visitandines according to Teresa of Avila's directives; the novice mistress was to help those she was directing to realize the importance of prayerful self-knowledge by "sometimes read[ing] to them chapters of the *Way of Perfection* of Teresa. I highly approve of having that book read to Novices, for it is very useful."[84]

Another area of shared principles of spiritual direction is that of individual and communal responsibilities. Jane directed her charges, as did Teresa, to pursue perfection by conforming "ourselves in all things to the community, and never to depart from it by our choice," while remembering simultaneously "that God draws, although in different ways," all who desire him.[85] "All the religious of the Visitation" were also called to the higher stages of prayer, because the community as a whole was so called, "but all were not led in the same way; each, according to the particular attraction of God in her, there being a difference so considerable and so great between them, that there are almost as many different degrees as there are souls that practise it."[86] Similar to Teresa, Jane advocated spiritual direction for all aspirants to spiritual perfection, telling Madame du Tertre that she "will receive a great deal of satisfaction" and "spiritual benefit" if she would only "be very open" with her Mother Superior and spiritual director (Mother Jeanne de Bréchard) "so that she may, by understanding your heart's weaknesses, bring to it appropriate remedies."[87] Years later Jane was still cajoling her into realizing the necessity of spiritual direction: "You would be happiest if you could be satisfied with the instruction of your good Mother" and to "never doubt that through her, His goodness will guide you safely. I am convinced that whoever gives up following the guidance of her Superior stops following that of God as well."[88]

Like Teresa, Jane insisted that every nun had had an apostolate of prayer. "We, who are called out of the world and its uproar, think not enough, if I mistake not, of the obligation we are under of tending to the perfection of our vocation, which, in substance, is nothing less than the total overthrow of the natural man and the union of our soul with its God."[89] Visitandine women sought "to unite ourselves to God by the entire, punctual, and exact observance of our Rule, Constitutions and all that concerns our little Institute," and specifically "to pray God for the people." Jane took this apostolate quite seriously and worried whether the community was fulfilling it properly: "I have thought myself bound to tell my Sisters the great distress in which this poor town is lying, for I have much fear that we are not careful enough in praying and calling upon God for the same, wherein, certainly, we shall be greatly responsible before God"[90]; indeed, "to pray for the public is one of the things for which we are gathered together, and I conjure you to do this carefully, for you obliged thereto by charity."[91]

If the women do not fulfill this apostolate of prayer, there is little justification for their vocation or for "God exempt[ing] us from the grievous trials those in the world are suffering," Jane pointedly argued. When they do fulfill it, however, society is profoundly indebted to them. "Why think you, my dear Sisters, has God drawn us out of the world to place us in religion? That we may serve him in holiness and justice all the days of our life; that we may pray Him for His people, for our good Christian brethren, for our dear neighbors who are in such suffering that it is intensely distressing to hear of all they are going through."[92] The women provided a sympathetic ear and heart in a society where the distressed had few opportunities to find such consolation. "One comes to tell us that all his relatives have died of pestilence, and that the skirmishing parties have ruined him. Another says: We know not how soon all our goods will be pledged," while a third confesses soldiers have killed his neighbor and "he know not when his life will be taken." Women come to the nuns who "are violated and weep their disaster; women are dishonored and their husbands killed; widows and orphans are oppressed." In exchange for being "free from those great evils which worldings suffer,"[93] the nuns must fulfill their role in society living their life of prayer for themselves and their neighbor. "Let us bear great compassion towards our neighbor, let us pray for him without ceasing. Let us weigh a hundred times a day, if possible, the benefits we receive from the hand of God" and then make sure we "put our talent to use." Jane's Visitandine directees "must labor in His vineyard to please Him and to receive wages, otherwise one is accounted useless."[94]

Jane's own contribution to spiritual direction, beyond her adaptation of many Teresan directives to her own order's particular situation, is perhaps best seen in her emphasis on simplicity and gentleness. In an increasingly complex and in some ways harsher society,[95] a call for simplicity and gentleness was much needed. "There is nothing which makes us more like God than simplicity; he who really possesses it is perfect," Jane preached at one of her conferences[96]; she repeated this dictum almost verbatim at numerous conferences and made it the chief topic in four other conferences.[97] "Perfect simplicity, my Daughters, consists in having but one single and sole pretension in all our actions, and that is of pleasing God in all things," Jane taught. The practice of such simplicity will in turn lead one "to see only the will of this great God in all things," to "open our faults" to sincere scrutiny, "to be truthful in our words," and, finally, "to live from day to day, without forethought or care for ourselves, but to do well at each moment."[98] The soul who lives by true simplicity "is no longer attentive to follow the lights of its self-love; it no longer listens to its persuasions and will no more look at its own inventions, which would seek self-esteem by great undertakings and by preeminent actions which may make us distinguished from the common. Such a soul enjoys peace ever tranquil," Jane continued, "it can say that it is free to rise above itself, by the possession of the Divine union." God himself uses "a certain holy simplicity" to teach "the souls to depend only on God, to love God only, to obey God only, and in the things of God, and not our own inclinations." The soul "ought never to turn" aside from simplicity, for it brings the soul to God and "what more desirable and better good is there than to repose wholly on God?"[99]

Gentleness, on the other hand, is not a virtue as much as it is an approach to life and to the spiritual. Time and again Jane directs her charges to "live in your house with perfect gentleness,"[100] to "imitate our gentle Savior and Master,"[101] to "try to become truly humble, gentle and simple,"[102] to approach difficulties "with great patience and gentleness,"[103] to practice "gentleness in our conversations,"[104] and so on. Everything a person does in pursuit of perfection must be done gently; especially when engaged in charitable works "we must have great gentleness."[105] Even when dealing with oneself, gentleness is required. Jane tells us in her commentary on the Visitandine Rule that she and Francis de Sales often discussed the necessity of a gentle approach. "If you have your soul in your hands," Francis told her, "and if you see that it has escaped from you, go after it and pick it up again. But, remember that you must take gently and softly" if you wish not to scare it off. "This is my counsel to you," Jane advised. "Look often, lest inclination wound it"; but if it does "then, very gently repair the disorder."[106]

In *Conferences* Jane directed her Sisters to pursue a similar approach when dealing with their passions. "Sisters, here is a little model of what we are to do, when, rowing peacefully in our little boat, we feel, without thinking of it, all our passions arise and cause a great storm within us." Rather than rage against the tempest itself, "we must gently draw near the shore" and thus "go gently along, without effort, and without yielding to our passions anything they wish, and by so doing, we shall arrive a little later in that divine port."[107] When writing to a fellow superior at Rumilly, Jane noted that "every day I notice that kindness, gentleness, and support as well as generosity, can do much for souls," and she advised her to change her harsh approach toward the women at Rumilly. "Be more and more careful that your corrections are not too harsh," Jane counseled, because "in the end, gentleness plays a large part in the way we govern."[108] Jane chastised another superior similarly: "Concerning the lack of gentleness that you show your Sisters when they bother you with trifles, you know that you must not behave in this way."[109] Novice mistresses were advised to guide their directees not "with affection, but cordially and gently," and to live in the "spirit of gentleness and trust, even as you bring others to do the same."[110] If those in charge would forget Jane's directive, she was quick to remind them, as she did with Marie-Aimee de Rabutin, superior at Thonon: "Do well what I advised you some time ago: help souls along very gently by word and deed and good example."[111] To another superior Jane wrote "I know there is no better way to succeed in leading souls" than to lead them "with a care that is genuine, loving, and gentle."[112] Jane also advised directees to respond to spiritual direction with gentleness. "Do everything you are taught in a spirit of gentleness and fidelity in order to reach the goal toward which you are being guided," Jane counseled, while she demanded the whole community "show a childlike trust and gentleness toward one another."[113]

Jane's emphasis on gentleness was quite deliberate and self-conscious; she believed it was the means by which she could lead many who were previously excluded from religious life and its accompanying spiritual direction into a community. Jane was particularly concerned with the physically infirm and tells us her realization came through personal experience.

> God permitted at the beginning of the Congregation, six weeks after it had been founded, that I should be attacked by great illnesses, without which it would have been very difficult to establish the Institute in the gentleness of rule in which it now is, and I said sometimes: My God! Thou art indeed provident and very merciful to treat me thus in order to accomplish more easily Thy designs, viz: that these houses should serve for the retreat of the infirm; and I myself inclined much more to the side of rigor and austerity, in which perhaps I corresponded more to nature than to grace.[114]

And while Jane saw her mission narrowly in terms of the infirm, the implications and repercussions of her decision to address the spiritual needs of the disabled were vast. By encouraging one group of people who previously had little opportunity to seek spiritual direction, she signaled the possibility of its availability to all. Gentleness is not a virtue we often consider to be capable of great things, but in this case Jane used it to help pry open a world often restricted to the religious elite. This is one of the reasons we call the seventeenth century a golden age of spiritual direction.

At the other end of the spectrum we find English recusant Mary Ward. She did not experience the general acceptance that Teresa and Jane did and, in particular, had to fight hard and long for her vision of religious life. Consequently, her spiritual advice is more combative[115] than gentle. It is still of prime importance, because it, too, helped direct women into previously restricted areas. Mary's own life was filled with conflict. After spending many difficult years trying to discern how and where she could best pursue perfection, "there happened a thing of such nature that I knew not, nor even did know, how to explain. It appeared wholly divine" and made it clear to her that she was "to leave what I loved so much and enjoyed with such sensible contentment, to expose myself to new labours, which then I saw to be very many, to incur the several censures of men, and the great opposition which on all sides would happen."[116] In practical terms this meant she must leave the "quiet and continual communication with God which strict enclosure afforded" and instead occupy herself with teaching children. Because traditionally "to teach children seemed then too much distraction, might be done by others, nor was it of that perfection and importance" of established religious orders.[117]

Mary Ward knew she had a difficult task before her. Not only did she have to accept the trials this new way to perfection would bring, particularly hierarchical resistance, but she must also pioneer new ways to guide others along the same path if she hoped to be successful. Clearly perceiving the ensuing struggle, her spiritual advice to those willing to engage in this new ministry with her were appropriately defensive. When Mary attempted to adopt Jesuit constitutions as the rule for her new group, everyone involved—the Jesuits in the area, her confessor, the local bishop, neighboring clergy—attacked the feasibility of the undertaking and insisted on other plans and rules. Mary stood firm: "There was no remedy but to refuse them."[118] Undeterred by the attacks, she persisted in her way. Mary enshrined her defensive position in her *Scheme of the Institute* by anticipating future attacks on her vision: "We hope and we humbly beg that neither the Bishop nor any one appointed to make the annual visitation shall have over us any other authority," Mary wrote, "that he may neither change nor add anything thereto, either with regard to our end or to the means by which it is to be attained."[119] In her notes from a retreat in 1616 we see her apply a defensive position to her own spiritual life: "I will endeavor that no sensible motions nor occurant accidents change easily by inward composition or external carriage, because freedom of mind and calmness of passions are so necessary both for my own profit in spirit and proceedings with others."[120]

Mary Ward's vision was bold. Consequently, her direction took an offensive as well as a defensive stance in the post-Reformation battle for souls. "As the sadly afflicted state of England, our native country, stands greatly in need of spiritual labourers," Mary observed, "it seems that the female sex also in its own measure, should and can in like manner undertake something more than ordinary in this same common spiritual necessity." Accordingly, "we propose to follow a mixed kind of life," such as that lived by the Virgin Mary and "Mary Magdalen, Martha, Praxedes, Prudentiana, Thecla, Cecilia, Lucy, and many other holy virgins and widows." The end of such a life was simply stated by Mary Ward: First, "to work constantly at the perfection of our own souls," and, second, "to promote or procure the salvation of our neighbor, by means of the education of girls, or by any other means that are congruous to the times."[121] In Mary's final plan for the Institute, the defensive nature of her group is explicitly stated; any woman who wishes to join the Society should remember that "she is a member of a Society founded primarily for this purpose: to strive for the defence and propagation of the faith and for the progress of souls in Christian life and doctrine."[122]

Given the daring nature of the plans for her community, it should come as no surprise to hear her both attack opposition and defend her Institute with vigor in conferences to her charges. In a series of spiritual instructions, Mary demolished the argument that women are incapable of attaining the same degree of spiritual perfection as men and actually directed women to such perfection. Mary recalled when "a father who loves you well" could not be persuaded to "think otherwise than that women are yet by nature full of fears and affections, more than men."[123] We have already recounted how the Father Minister himself disparaged the Institute's initial accomplishments by claiming their initial success was only the result of first fervor. Mary's advice to the women in the face of such discouraging attitudes reveals the starting point of her spiritual direction: Spiritually people are genderless. "'*Veritas Domini manet in aeterum*': the verity of the Lord remains for ever," Mary began. "It is not *veritas hominum,* the verity of men, nor the verity of women, but *veritas Domini,* and this verity women may have as well as men. If we fail, it is for want of this verity, and not because we are women." The social roles and relationships between a woman and man may vary, "but in all other things, wherein are we so inferior to other creatures that they should term us 'but women'? As if we were in all things inferior to some other creation, which I suppose to be men! Which, I dare be bold to say, is a lie and, with respect to the good Father, may say it is an error. I would to God that all men understood this verity, that women, if they will, may be perfect, and if they would not make us believe we can do nothing and that we are 'but women,' we might do great matters."[124]

Mary did not stop there but continued to direct women in the ways best suited to achieve their desired happiness.

> Now you are to understand how you are to attain this perfection. By learning? No, though learning be a good means, because it giveth knowledge. Yet you see many learned men who are not perfect because they practise not what they know, nor perform what they preach. But to attain perfection, knowledge of verity is necessary, to love it and to effect it. That you may not err, I beseech you all to understand and note well wherefore you are to seek this knowledge. Not for the content and satisfaction it bringeth, though it be exceedingly great, but for the end it bringeth you, to which is God. Seek it for him, who is Verity. Then you will be happy and able to profit yourselves and others.[125]

This last point is key to Mary's direction. Actions are neutral unless they are done for the right reason, and if done with proper intention then happiness follows: "Remember then that [God] be the end of all your actions and therein you will find great satisfaction and think all things easy and possible."[126] We find this principle behind some of Mary's earliest spiritual directives[127]: "We ought to work and suffer for God, and for the rest let Him make use of us, according to His good pleasure, for the fulfillment of His most Holy will should be our sole wish and only desire."[128] If the women approach their work this way, they will cease to worry about the work itself: "What ever falls to thee to do, that perform as much as thou canst faithfully and diligently, but be not too careful as to how it may turn out, nor whether it will be hazardous or not, but commit it to the good God."[129] Moreover, any burden the work entails is easy to bear, for the end is worth it. Thus the women should "be ashamed to say that anything appears hard to thee in the service of God, for to those who love all is light," and one cannot really "satisfy thyself with nothing which is less than God."[130]

The Institute of the Blessed Virgin Mary, as Mary's group was called, was judged to be too innovative and contrary to the expectations of the day, particularly in its attempt to identify so closely with the Jesuits. The Jesuits resisted formal association with the women, "for it is clear enough that the Jesuit Fathers are expressly forbidden by the precepts of their own rule to involve themselves or meddle with the government of any women whatsoever."[131] In 1631 the Institute was formally suppressed in a papal bull, and Mary was even imprisoned for a short time. True to her own directives, she prayed that "we may more firmly and constantly persevere in own intention"[132] and thus eventually got papal permission to start over again with a new rule. In 1703 the Institute was granted full canonical status as a religious congregation.[133] A less overt reason for the Institute's problems may have been Mary herself and her aggressive spiritual direction; it was even implied in written attacks on the Institute that Mary and her direction were the main problems. However, the first known criticism, *Concerning certain English Virgins,* lists many reasons why it was problematic, but does not mention spiritual direction per se. Instead, it claims the group was dangerous because the members were "not keeping enclosure; of going about too freely; of aspiring to apostolic work, even among heretics; of filching vocations from existing convents"; and "of being closely connected with the Jesuits."[134] Still, the *Memorial of the English Clergy to the Holy See* did imply resentment and jealousy over the women's role as spiritual directors to those they worked with: "The aforesaid presume and arrogate to themselves authority to speak about spiritual things before grave men and even sometimes when priests are present, to hold exhortation in an assembly of Catholics and to usurp ecclesiastical offices of that kind, as is manifest by daily custom."[135] If Mary's spiritual direction was indeed an insurmountable problem, then in this area, as in so many, many other ways, Mary Ward's story is not typical.[136]

In some ways, Marie of the Incarnation was of similar ilk as Mary Ward; she also shared much with Teresa of Avila and Jane Frances de Chantal. Born Marie Guyart in 1599 in Tours, she married, had a child, and was widowed all by age nineteen. She tells us that "his Divine Majesty delivered the coup de grâce" through a mystical experience,[137] and thus she began her spiritual quest in earnest. During this early stage Marie read Teresa's *Life* and *Way of Perfection,* which "consoled me" and then became, along with Francis de Sales's *Introduction to a Devout Life,* the foundation of her spirituality.[138] As years passed and her son grew to adulthood she contemplated joining the newly established Teresan reform Carmelites. "[F]or my part, I dearly loved this holy order,"[139] but in 1631 she decided instead to join the Ursulines. Here she was quickly "assigned to the novitiate to assist the Mistress of Novices" where her "task was to teach Christian doctrine in order to form the novices into capable members" of the community.[140] As we read Marie's own assessment of how she guided the novices under her charge, it is easy to identify the many spiritual trends whirling around in seventeenth-century French women's religious quarters.

Before beginning my instruction I would read something from the little *Catechism of the Council of Trent* or that of Cardinal Bellarmine, although I did this rather seldom. When, after speaking of some article of faith, I would turn to questions of morality. I was amazed at the number of relevant passages of Holy Scripture that would come to mind. I could not remain silent, for I had to obey the Spirit who possessed me. All my life I had had a great love for the salvation of souls, but since the kiss of the most holy Virgin I have had a consuming fire for this. Since I could not travel through the world as I would

have wished to win souls, I did what I could in the novitiate, adjusting myself
to each one's individual capability. There are some very fine souls there. They
would urge me more and more to continue my instruction. God also desired
this of me, and I experienced interiorly that it was the Holy Spirit who had
given me the key to the treasures of the Sacred Word Incarnate.[141]

Soon Marie believed that "his Divine Majesty wanted to put me in a new
state," one where she "was wholly dedicated to zeal for his glory so that he
would be known, loved, and adored by all peoples."[142] The answer to where
that state was came when a Father Poncet sent her "a picture of Mother Anne
de Saint-Barthelémy, a Spaniard, in which Our Lord was depicted with his
hand pointing toward Flanders, inviting her to go to serve him there." As
Ana of San Bartolomé responded to the call to establish Teresan houses, so
too should Marie respond to her personal call. "'I'm sending you this picture,' he wrote, 'to urge you to go to serve God in New France.' I was
amazed at this invitation," which, Marie confessed, "was like a spur which
activated still more powerfully the fire for the salvation of souls which consumed me."[143] Thus she was convinced to serve in Canada. In 1639 she arrived there to begin her apostolate to found and guide the first women's
monastery in the territory.

Actually, Marie's ministry in spiritual direction began years before, while
she was still a young widow trying to support herself and her son while living
with family. She tells us that "I sometimes spent time with a group of men,
servants of my brother-in-law, sitting alone at table with twenty or so of these
simple people (depending on the number who gathered together, coming in
from the country) to have the opportunity of instructing them about their salvation." There was apparently no resistance or offense taken that a woman
would so do, for "they would give me very simply and intimately an account
of their actions, pointing out to each other the faults they had committed
when, through forgetfulness, someone had neglected to mention something."
They quickly and willingly learned to render the traditional obedience directees give to a director, being "as submissive to me as children. I even made
them get out of bed if they had retired without praying to God." As director
it was Marie who "brought them together from time to time to speak of God
and teach them how to keep the commandments" and even "reprimanded
them openly." She saw her role as director as all encompassing, not restricted
solely to the spiritual. The whole person must be guided toward perfection,
and so she did not turn them away when "they came to me for help in all their
needs, above all in their sickness, or to have me make peace with my brother
[-in-law] when they had displeased him. I felt impelled to do all this."[144]

This approach served Marie well, for it was adopted by the Ursulines.
From "the day following our arrival" in Quebec, she believed it was her duty

to clothe, feed, educate, and spiritually train "both Indian girls and the daughters of the French who trade in this country."[145] Marie realized from the very beginning that the spiritual training of the Indians had to start with their physical care, even with basic hygiene: "Yet all of this never disgusted us; on the contrary, each one considered it a privilege to clean up" the natives who came to them for training.[146] Some of the nuns in the settlement worked solely in the hospital, and since "the hospital being entirely open and the good things done there seen by everybody," their work was often "rightly praised" by all.[147] The work of education and spiritual direction, however, took place in "our cloister [which] hides everything." Here "the mistresses work hard on [the natives'] education, and they teach them, sometimes with in a single year, to read, to write, to count, along with their prayers, Christian morality, and everything a young girl ought to know."[148] Marie ardently believed that "one must undertake everything for the service of God and the salvation of our neighbor," and thus tells us that "at the age of fifty I am beginning the study of a new language," that of the Hurons, after previously having mastered "that of the Algonquins and Montagnais."[149] It was "the urgent desire I had to teach them [which] led me to be the first to embark on this," Marie confessed, and "in a short time I knew enough to be able to instruct our beloved converts in everything necessary for their salvation."[150] And to Marie, the converts were indeed beloved: "I carry them all in my heart and try very gently through my prayers to win them for heaven," she confided. "There is always in my soul a constant desire to give my life for their salvation."[151] It is because of such dedication and its resultant effectiveness that Marie could unequivocally state "that were it not for the Ursulines, their salvation would be in constant danger."[152]

Marie of the Incarnation's spiritual direction intertwined with her wholehearted embrace of every trial that one meets in life. When Mother Marie-Gillette Roland asked Marie what she believed was the way to perfection, Marie explained why missionary work was such an effective way to reach such a goal. "I see that those people whom one believed to have had some perfection when they were in France are now, in their own eyes and in the eyes of others, very imperfect" in their apostolate in New France, because "the more they work the more they discover imperfections in themselves. The reason is that the spirit of the new Church has such a profound purity that the slightest imperfection is incompatible with it." Marie sees "this primitive Christianity as a kind of purgatory in which, as these souls cherished by God are purified, they share in the communications of his divine Majesty. This very same thing, I say, happens here. This hidden spirit, which is none other than the spirit of Jesus Christ and of the Gospel, gives the purified soul a certain knowledge of itself, establishing it in an interior life that brings it closer of its resemblance of Christ."[153] Trials were but purification to be embraced willingly because of this end result. Marie's own purification came chiefly through her work; in fact, she identified for us the trial she considered hardest to bear: "The most agonizing [cross] that I have ever suffered during fifteen years I have lived in this new Church—or during my whole life—is that pertaining to our converts: Algonquins, Montagnais, Hurons— who for the last ten years have been the prey of their enemies. I could never express the anguish I have suffered on these occasions."[154] But whatever the trial was for each individual, it was always an opportunity: "[T]he time for proving your fidelity to Him who has been so merciful to you," Marie told one of her directees, comes when one begins "the actual practice of those virtues which you have seen and experienced during your prayer. You must realize, my dear daughter, that you have a very big job to do and that consequently you have no time to lose."[155]

Louise de Marillac also exemplifies the best of this golden age of spiritual direction. Just as Jane de Chantal's work as a spiritual director has been for centuries overshadowed by her male partner Francis de Sales, Louise de Marillac has only recently emerged as a spiritual leader independent of her coworker Vincent de Paul. Like so many women leaders in the past, much of Louise's greatness lies in her ability to direct others along the same path toward perfection that she pursued. And, again, like many before her, she build on the existing tradition of women's spiritual direction and improvised when needed. Although she was apparently happily married to Antoine Le Gras with one son, in 1623 Louise made "a vow of widowhood should He call my husband to Himself." Her opportunity to fulfill her vow came two years later with Antoine's death. Around the time of her vow she also received inner assurance that her lifelong desire to "help my neighbor" would be realized "in a small community where others would do the same" with "much coming and going."[156] Such a life was actualized when in 1633 Louise, with Vincent de Paul's approval, vowed to dedicate herself to the development of the newly founded Company of the Daughters of Charity. Together Vincent and Louise succeeded in finally creating an active religious order for women without enclosure and with hierarchical approval.

Theirs was a fruitful partnership. Vincent guided Louise and she in turn guided the women who joined the Company[157] in a combination of approaches that was uniquely her own.[158] Mary Ward directed her directees to defend any attacks on their spiritual work vigorously and explicitly; Louise did so too, but perhaps with more subtlety and tact. When the Daughters' work in Angers was "being refused by the administrators" of the city and Louise realized "that these gentlemen are not really happy with our sisters," she wrote Guy Lasnier, the vicar-general of Angers and protector of the Daughters at the Hospital of St. Jean, and begged him "most humbly to take the trouble to let me know if you know some way to bring about the order in this matter." She even wrote the men involved "in order to allow them the opportunity to air any complaints they might have."[159] Marie of the Incarnation directed her directees to embrace physical work and trials as a means of purification; likewise, Louise reminded her Daughters that when faced with a trial they "must accept this trial as coming from Divine Providence as our share in the Cross of Our Lord and as an opportunity He has given you to enable all of us to follow Him."[160] But Louise also urged others to actively assist those women undergoing trials: "The work of our poor sisters here is almost unbelievable; not so much because of the great effort involved as on account of the natural repugnance one has for this type of work. For that reason it is most fitting to help them, to encourage them, and to make known what they are accomplishing and its value in the sight of God. It is also fitting to help them by our prayers."[161]

Jane de Chantal tried to approach both her directees and their problems with gentleness; Louise used it in the same manner but also employed it to accomplish the work of salvation. "Gentleness, cordiality, and forbearance must be the practices of the Daughters of Charity," Louise instructed the Daughters at Richelieu.[162] "As for your conduct toward the sick, may you never take the attitude of merely getting the task done. You must show them affection; serving them from the heart—inquiring of them what they might need; speaking to them gently," Louise told the Daughters at Montreuil. "Great gentleness and cordiality are necessary in order to win over these people."[163] Gentleness sometimes was even needed to win over the hierarchy in order to get their apostolic work done; Louise advised Julienne Loret and the Daughters at Chars to deal gently with a new pastor who "is a bit brusque": "The only way to deal with him is by gentleness, not by arguing and by doing what he asks when you can. When you cannot carry out his request, you must explain your reasons to him gently and humbly."[164]

Last, Teresa of Avila accepted spiritual direction in any form and from any person; Louise used Teresa's approach as her model and instilled it in the foundation of her order's spirituality. "Recall, my dear Sister, the great Saint Teresa," Louise explained to her close associate Jeanne Lepintre, "who was much busier than we and charged with affairs of much greater importance and who often needed advice. Although the advisors she wanted were absent, she was so simple and humble that she freely sought advice from those whom Providence sent as directors. She listened to them as if God were speaking to her."[165] Louise had adopted Teresa's attitude early in her spiritual life, for she confessed that the acceptance of spiritual direction from any source was part of her conversion experience. Prior to the experience "I also doubted my capacity to break the attachment I had for my director"; afterward she felt assured "that God would give me one whom He seemed to show me. It was repugnant to me to accept him; nevertheless, I acquiesced."[166] When directing others she tried to guide them toward such acceptance, knowing full well "the suffering you experience because of your reluctance to open your heart and because of your belief that no one is suited to direct you." When directing such a person Louise argued that it was precisely when we think that "human beings fail us" that "God reveals Himself more abundantly to us,"[167] either directly or through a person we least expect. Louise was also rather insistent that the first person the women must look to for direction was the Sister Servant (the community's superior), for "I consider this absolutely essential in order to maintain peace." It is the Sister Servant who "is the one who directs the others albeit in the spiritual more than in the temporal domain."[168] Louise was well aware "that direction is a gift which must be obtained through patience,"[169] but she was even more aware of how difficult and humbling the role of spiritual director was. Once when faced with numerous spiritual, physical, and emotional dilemmas among her directees she confessed to Vincent in frustration: "I am a bit overwhelmed by all the spiritual and emotional problems faced by the greater number of our Sisters. I assure you, Monsieur, that my inability to help these good girls reach perfection is a subject of humiliation for me before God and the world."[170]

In her direction Louise dealt with another situation, one not explicitly treated in the direction of the previously discussed women directors: change. This was indeed a significant addition, because it indicates an awareness of the need to adapt traditional monastic direction to the more activity-prone religious of the modern era. In the record of her conversion experience, Louise mentions her reaction to change twice. First, when God revealed to her that she must accept a new director, she found, as already noted, it "repugnant to me to accept him" and resisted wholeheartedly. The reason why she reacted this way was because "it seemed to me that I did not yet have to make this change."[171] It was not a specific person that repelled her, for she did not know him; it was the change. Second, when she received assurance that she eventually would be able to join a small religious community "where I could help my neighbor," she still had difficulty accepting the mystical communication because "I did not understand how this would be possible since there was to be much coming and going"[172] within the community. Her main acquaintance with religious life to date consisted of stable, enclosed communities; she had difficulty envisioning a religious life that embraced constant change in place, task, and personnel: "much coming and going."

In many ways this was a dilemma of the new age. Louise and her Daughters had to abandon all resistance to change. Without full acceptance of change their apostolate would never succeed. This dilemma was, of course, not unique to seventeenth-century France or to Louise and the Daughters of

Charity. Change is a universal reality in every society, and each society must develop its own way of dealing with change. In seventeenth-century France, society had to face the reality of the poor in unprecedented ways; traditional sources for treating the problem—monastic houses, parishes, personal charity—were inadequate. It was in large part for this reason that the Daughters of Charity were founded, to meet the needs of this pressing, ever-changing dilemma.[173] As Louise informed Cecile Agnes, every Daughter of Charity must acknowledge that "changes are always difficult, and that it takes time to learn new ways of serving the poor skillfully and well."[174] The Daughters needed to be guided carefully through this transition from stable communities to ones flexible enough to respond at a moment's notice to a new situation. Change "is partially responsible for creating a certain flightiness among them. However, I think that it is an evil that is almost essential for the government of the Company."[175] Louise believed the success of the Daughters rested on their ability to embrace change inwardly on a spiritual level, for without that acceptance external physical changes would always chafe and be resisted. It was with this in mind that Louise explained her reasons for insisting on change at a conference given to the Daughters.

> The first reason that obliges the Daughters of Charity to accept changes of place, persons and duties is the respect that they owe to the example of the Son of God who acted in this way. The second is that such changes can and must occur. If they are not accepted, we shall never enjoy the peace of soul that is essential if we are to please God and to accomplish His holy will. . . . If the sisters are not willing to accept changes in all these circumstances, they are in danger of committing many faults. . . . Another evil is the disedification that we would give to our neighbor. Moreover we would find ourselves in the impossibility of faithfully practicing our Rules and we would be in danger of losing our vocation. There are an infinite number of other evils which would befall us but which are too lengthy to mention here.[176]

In Louise's defense of change was a profound reality her contemporaries needed to hear. Happiness was not to be associated with stagnation but with change.

Chapter 8

⁓

Spiritual Direction
among the Protestants

For various reasons spiritual direction since the Reformation has been associated more closely with the Roman church than with Protestant traditions. To a large degree this assumption is valid. The Reformation era saw the role of the spiritual director often wedded to that of the confessor; the Protestant deemphasis on sacerdotal powers, particularly those exercised in confession, led many Protestants to grant less attention to spiritual direction. On the eve of the Reformation spiritual direction was found mainly within religious orders; since Protestant churches disavowed monasticism, they had de facto eliminated spiritual direction's chief locus of operation. Jesuits advocated spiritual direction; Protestants were bitterly anti-Jesuitical. Sixteenth-century spiritual direction literature in the Roman tradition talked at length about the stages and levels of development in the spiritual life; Protestants avoided references in their literature that reinforced privileged class terminology employed by the clerical estate. Finally, Romans retained and continued to produce great spiritual masterpieces that were the basis of their spiritual direction; Protestants immersed themselves in Scripture, theology, and organization, and only rarely in spirituality.

This is not to say that mainstream Protestants condemned or even ignored spiritual direction. To the contrary, many of the first reformers actively engaged in it themselves. In his *On the True Cure of Souls and the Right Kind of Shepherd,* Martin Bucer included spiritual direction in his list of proper duties pastors must fulfill. For Bucer, the good priest was "to lead back those who have drawn away; to seek amendment of life in those who fall into sin" and to guide "them forward in all good."[1] Zwingli reduced sacramental confession to spiritual direction, "a consultation in which we

receive from him whom God has appointed . . . advice as to how we can secure peace of mind."[2] Calvin was the personal spiritual director of many,[3]
as was John Knox. With Luther, the historian is most fortunate, for he left
posterity with volumes of letters of spiritual guidance. Indeed, by the end
of the sixteenth century there were already five collections of his letters of
direction completed, and almost every generation, including our own, has
produced a new collection.[4] These letters are in addition to the spiritual direction recorded by those privy to his private discussions published as *Table
Talk*.[5] "Your preachers will tell you enough of what is to be said about the
Sacraments and external things, about eating and drinking, dress and conduct," Luther wrote to some Christians in Livonia in a letter that typified
his approach to spiritual direction. "All these external things will take care
of themselves in Christian freedom," he continued, because "everything depends on faith in Christ and love to one's neighbor."[6] Still, the descendants
of these first reformers did not contribute much to spiritual direction literature[7] or nourish in their churches a strong tradition of direction.

The same cannot be said about the radical reformers. Within many of
these sects, particularly those that grew out of the Anglican tradition, there
arose a visible and articulate tradition of spiritual direction and, most significant to us here, a tradition of women spiritual directors. This is especially
true of Puritans, Quakers, and Methodists.[8] Puritans, with their emphasis
on conscience, gave birth to a small body of literature that argued for the necessity of spiritual direction,[9] and we have some witnesses to women so directing consciences in conflict. For example, Richard Baxter tells us his wife
Margaret "was better at resolving a case of conscience than most divines that
ever I knew in all my life." He often told his wife of difficulties involving
problems "about restitution, some about injuries, some about references,
some about vows, some about marriage, promises, and many such like; and
she would lay all the circumstances presently together, compare them, and
give me a more exact resolution than I could do."[10]

While Puritans did much to promote the woman as the center of domestic well-being, as a group they did not foster a sense of spiritual equality
of the sexes. Historian Peter Lake has argued convincingly, however, that
many an English Puritan woman discovered that "complete subjection to,
and possession by, her God enabled her, if not to resist, then to circumvent,
the usual constraints of female existence."[11] If women did not achieve the
recognition of spiritual equality, they at least achieved access to certain spiritual activities, including spiritual direction. Lake's case in point is Jane Ratcliffe, whose life is recorded by her pastor, John Ley. Ley tells us that after
the death of her husband, Jane began limiting her social life to women, "for
she most scrupulously shunned the presence of men, especially of ministers,
in all the sacred services which in private she performed."[12] Very well edu-

cated, "her reading of scripture [was] very solid and substantial," and she possessed such "a deep insight into the sacred text" that Ley "seriously desired her to write down her observations on the bible."[13] Despite her gift, she decided not "to teach any but her children and servants"[14] but instead to engage "in spiritual intercourse which surpasseth the pleasure of all sensual enjoyment." Eventually a group of close religious for whom "she entertained with most unfeigned affection"[15] formed around her in order to benefit from her spiritual direction. Her guidance was given with "such discretion and moderation that none took her reproofs for reproaches but so accepted of them that (if they gave her not cause by their reformation to think better of them) they showed no signs that for her reprehension they thought the worse of her."[16]

In America there was a similar situation. First-generation Puritan women like Anne Hutchinson who openly and explicitly ascertained their spiritual equality were deemed "unfit for our society"[17] by the male hierarchy in power. Still, we find the Puritans acknowledging the principle behind women spiritual directors at least somewhat, for John Cotton, a witness at Anne Hutchinson's trial, made it the basis of his argument in his *Singing of Psalms a Gospel-Ordinance*. Here Cotton maintained that there was "ground sufficient to justify the lawful practice of women singing together with men the praises of the Lord," thus signifying, according to historian Rosemary Skinner Keller, "women's emerging role" in Puritanism as "bearers of religious piety" and "evangelists throughout the churches."[18]

By the second generation[19] American Puritan women were fulfilling the role of spiritual directors, although neither their contemporaries nor ours identify them as such. In her introductory remarks to Puritan Sarah Whipple Goodhue's *Valedictory*, historian Martha Tomhave Blauvelt-Keller admits that Goodhue's testimony illustrates that "one minimal evangelical role open to women in the late seventeenth century was preaching within the private circle."[20] When we read Goodhue's *Valedictory*, however, it is clear that what they called private preaching is indistinguishable from what we have established here as spiritual direction. Sarah Goodhue's testimony, written days before her death, is a summation of her spiritual advice to the "Brothers and Sisters All" of her "private society, to which while here I did belong"; it is of the same nature as the spiritual guidance given in monasteries throughout Europe between novice mistresses and novices. Sarah encouraged them to "be not discouraged, but be strong in repenting, faith and prayers, to pray each for another, and with one another," and to "think not a few hours time in your approaches to God mispent; but consider seriously with yourselves to what end God lent to you any time at all." She then addressed her "children, neighbors and friends," telling them that "I do not only counsel you, but in the fear of the Lord I charge you all, to read God's word, and, pray unto the Lord that he would be pleased to give your heart and wisdom to improve the great and many privileges that the Lord is at present pleased to afford unto you."[21]

Puritans eventually acknowledged that although "Women full of Good Works, have illumined all Ages of the World"[22] who "have not been inferior to [saints] for the Piety of their Lives, and yet have passed silently thro' the World, without so much as an Epitaph at their Deaths,"[23] other sects made these facts central to their beliefs from their origin. Chief among these groups were the Quakers, Shakers and Methodists. Before continuing our survey, however, there are two caveats concerning sectarian women spiritual directors of the period that should be noted. First, Puritans, Quakers, Shakers, and Methodists did not identify the function women were fulfilling as spiritual direction. The spiritual director was not a recognized term in these societies; it is not found in any of their sources. Indeed, one wonders how much confusion would have dissipated and how many arguments been avoided if the activity many women were engaged in had been properly identified as spiritual direction. Instead, generations of these groups had to defend themselves against accusations that their women were preaching, not keeping silence in the churches, or simply infringing on areas of ministry traditionally preserved for men. These women were frequently judged to be usurpers of male authority in direct disobedience to Paul's mandates in I Timothy 2:11–12. In reality what many of these women were engaged in was the ministry they had participated in since the time of Christ, spiritual direction.

As we have also seen, the methods and ways of spiritual direction have changed repeatedly throughout the centuries. During Jesus' lifetime, spiritual direction was often as simple as directing a person to Jesus himself, as the Samaritan woman did. The early Middle Ages frequently saw the spiritual direction of people by bringing them face to face with the way to perfection through education and through example; Hilda of Whitby, Radegund, and Leoba practiced these approaches. Catherine of Siena offered her increasingly literate Renaissance contemporaries spiritual direction through letters that emphasized individual and communal responsibilities, free will, and self-knowledge, while Teresa of Avila presented a most incisive analysis of the interior life in a manner accessible to the uneducated layperson as completely as to the educated religious. The women of these modern sects were simply doing what many of their Christian predecessors had done, adapting spiritual direction to the needs of the people. Quaker women's testimony to the Inner Light and their facilitation of Women's Meetings were ways they consciously employed to guide others to "see the unity and agreement of our doctrine and testimony with the testimony of Jesus."[24] Methodist women chose exhortations and expounding to direct those present in their bands and classes to be "by faith brought into full liberty."[25]

During the Great Awakening in colonial America, women spiritual directors traversed the land to lead the masses to the Truth by "weeping and crying in a most extraordinary manner."[26]

While the methods these women used were occasionally different from the past, what they were doing was not. The women were directing people toward their spiritual end, just as so many women before them had done. This does not mean that they were in fact not preaching, not teaching, or not prophesying, as their contemporaries sometimes complained; many women were. Certainly, for example, many evangelical women leaders during the Great Awakening were called and called themselves preachers.[27] But the line between ministries is not always clear. It is my contention that the vast number of Protestant women, particularly among the Quakers and Methodists, who were identified as religious leaders were in fact following those within the revered tradition of women spiritual directors.

This brings us to the second caveat. Because this is so, then historiography that claims the position of women within these sects is without precedent or parallel is overstated.[28] Surely their sustained visibility was a key development in women's religious history. Just as surely it was not wholly unique. It was part of a long tradition. As historian Anne Laurence concludes without even taking into consideration the history of women spiritual directors, "despite the prominence of women in the sects" of seventeenth-century England, their position "was no different from that of women in the Church of England or, indeed, in the Roman Catholic Church."[29] When one takes the tradition of spiritual direction into account, Laurence's conclusion becomes even more persuasive. The role of women in many dissenting sects of the modern era were but variations of the role already established and being played out in the Roman church.

The sources indicate that the Quakers, probably the sect most well known for its support of women's active participation,[30] based their support of women's spiritual equality on the scriptural passage so popular among apologists for women's parity, the creation of man and woman in God's image and likeness (Gen 1:25–26).[31] Both George Fox, founder and organizer of Quakerism,[32] and Margaret Askew Fell Fox, often called the mother of Quakerism,[33] wrote powerfully and persuasively on the subject. "Let me lay down how God Himself manifested His will and mind concerning women and unto women," Margaret Fell began her famous pamphlet, *Women's Speaking Justified, Proved and Allowed of by the Scriptures.*[34] "First, when 'God created man in His own image, in the image of God He created him; male and female He created them. And God blessed them, and God said to them, 'Be fruitful and multiple. . . . ' Here God joins them together in His own image, and makes no such distinctions and differences as men do, for though they be weak, He is strong."[35]

Employing a rhetoric of femininity reminiscent of Teresa of Avila and sixteenth-century Spanish women, Margaret incorporated the image of woman as weak into her presentation but then turned the argument on its head by claiming this woman's weakness was precisely why she must be allowed to speak. After quoting Genesis 3:15, "I will put enmity between you and the woman, and between your seed and her seed; he shall bruise your head, and you shall bruise his heel," Margaret continued her argument: "Let this word of the Lord, which was from the beginning, stop the mouths of all that oppose women's speaking in the power of the Lord, for He has put enmity between the woman and the serpent, and if the seed of the woman speak not, the seed of the serpent speaks."[36] This fact of creation should "serve to stop that opposing spirit that would limit the power and Spirit of the Lord Jesus, whose Spirit is pored upon all flesh, both sons and daughters, now in His resurrection—since that the Lord God in creation, when He made man in His own image, He made them male and female."[37]

George Fox presented the same text as the basis for spiritual equality but pushed it even further by reasoning that women are not subordinate to their husbands by virtue of their equal creation: "For man and woman were helps meet in the image of God, and in righteousness and holiness, in the dominion, before they fell; but after the fall in the transgression, the man was to rule over his wife; but in the restoration by Christ, into the image of God, and his righteousness and holiness again, in that they are helps meet, man and woman, as they were before the fall."[38] By the mid-1670s reliance on this text was common among Quakers, and it found its way into a letter probably sent from Sarah (daughter of Margaret) Fell's Swarthmore Women's Meeting to "*the Dispersed abroad, among the Women's meetings every where.*" It began with a reminder: "So here is the blessed Image of the living God, restored againe, in which he made them male and female in the beginning: and in this his own Image God blessed them both, and said unto them increase and multiply." The letter continues to repeat the basic scriptural exegesis found in Margaret's pamphlet: "And in this dominion and power, the Lord God is establishing his own seed, in the male and female over the head of the serpent, and over his seed, and power. And he makes no difference in the seed, between male and female, as Christ saith, that he which made them in the beginning made them male and female."[39]

All three authors also bolstered their defense by calling upon the examples of biblical women, women already identified in this study as spiritual directors. "And they returned from the Sepulchre and told all these things unto the eleven, and to all the rest. It was Mary Magdalen, and Joanna, and Mary the Mother of James, and other women that were with them, which told these things unto the Apostles," Sarah Fell wrote. "Soe here the Lord Jesus Christ, sends his first message of his resurrection by women unto his own disciples: And they were all faithfull unto him, and did his message, and yet they could hardly be believed."[40] Years later, while clerk for the Lancashire Women's Meeting, Sarah wrote in a similar vein: "They were all gathered together in an upper room" when the Spirit came to the disciples "with the women and Mary the mother of Jesus" in such a manner that all were filled. "Therefore let all mouths be stopped, which would limit the spirit of the Lord God in male or female, which he hath not limited; but the Lord hath regard unto and takes notice of the women, and despises them not."[41] Margaret Fell used the example of the women in a much more confrontational way:

> Mark this, you that despise and oppose the message of the Lord God that He sends by women: what (would have) become (of) the redemption of the whole body of mankind, if it had not cause to believe the message that the Lord Jesus sent by these women, of and concerning His resurrection? If these women had thus out of their tenderness and bowels of love (who had received mercy, grace, forgiveness of sins, virtue, and healing from Him, (of) which many men also had received the like), if their hearts had not so united and knit unto Him in love that they could not depart as the men did, but sat watching, waiting, and weeping about the sepulchre until the time of His resurrection, and so were ready to carry His message, as is manifested, how else should His disciples have known, who were not there?[42]

Margaret also called upon the witness of the Samaritan woman[43] and Martha, both of whom manifested "true and saving faith, which for at that day believed so on Him"; the "many women who followed Jesus from Galilee, ministering unto Him"; Mary, Martha's sister; Mary Magdalen[44]; Joanna; Susanna; and "the daughter of Jerusalem" who wept for him.[45] Likewise, George Fox emphasized the role of women who "went first to declare the Resurrection out of death, out of the grave. Now, they said, 'certain of our company came and told us he was risen.' Certain *women* they were, disciples, learners and followers of Christ. This seemed as idle tales, but when they came into the belief of it, male and female believed: so both are one in Christ Jesus, and all can praise God together."[46]

This somewhat detailed discussion of Quaker scriptural understanding of women before discussion of our main focus of spiritual direction corresponds to the historical reality of seventeenth-century English Quaker women. The Clarendon Code of 1662 and the Conventicle Act of 1664 introduced a period of extreme scrutiny and persecution of the Quakers in England and continued until the Toleration Act of 1689 allowed them more freedom. The Quakers' legal status was not the only thing to change during these three decades, for the Friends responded to the external pressure by toning down their more ecstatic, radical behavior, constructing a more rational theology to contain their visionary insights, and organizing a somewhat tightly knit structure to control their enthusiasms.

Part of that response was the establishment of the Meeting system, an elaborate organization of male local, regional, and general meetings and a corresponding organization of meetings for women. The creation of the Women's Meeting was one of the more controversial developments of the period, both within Quakerism itself and in the larger society,[47] but it was most fortuitous for women. It was here that women's leadership roles, including spiritual direction, was institutionalized. Quaker women now had a sanctioned place and time where spiritual direction was expected to be given and received by women and to women. There were, of course, many other functions the meetings were expected to fulfill,[48] but given the intensity of the early Quakers' reliance on what they called the Inner Light,[49] spiritual direction to guide the individual's interpretation and understanding of that Light was chief among them. Spiritual direction was desirable both for enlightenment and control. As historian Margaret Hope Bacon concludes, "Since the Friends had no pastors, the principal concern of those meetings was the pastoral care of members. The women's meeting from the first played an active and necessary role in this work."[50] It was at meetings that women shared their personal spiritual experiences, even if it was difficult to reveal such intimate details at first. Mary Neale tells us that her first revelations "were in great fear, and under a feeling that my natural inclination would not lead me into such exposure, for I shrunk from it exceedingly; and often have I hesitated, and felt such a reluctance to it, that I have suffered the meeting to break up without my having made the sacrifice." Even though she "had opened my mouth in private," she was not satisfied until her spiritual experiences were submitted to the scrutiny of the women's meeting. When she finally did, she "had sweet consolation in coming into obedience; and after a while was surprised to find, that although I stood up in meetings expecting only to utter a *little* matter, more passed through me. I scarcely knew how." As thus "the gift grew" and she learned the ways of sanctification, her turn to guide others along the same path came, and she "was occasionally induced to invite others to the needful acquaintance with Him who came to redeem us from inquity."[51]

Clearly, it was at the meetings that women were guided by other women along the Quaker path to perfection. "Let the women likewise of every monthly meeting, meet together to wait upon the Lord, and to hearken what the Lord will say unto them, and to know his mind, and will, and be ready to obey," Sarah Fell writes in her *Letter to the Dispersed Abroad, among the Women's meetings every where.* "If there be any that walks disorderly, as doth not become the Gospell, or lightly, or wantonly, or that is not of a good reporte: Then send to them, as you are ordered by the power of God in the meeting (which is the authority of it) to Admonish, and exhort them, and to bring them to Judge, and Condemn, what hath been by them done or acted contrary to the truth."[52] All Quaker women must remember that "here in the power and spirit, of the Lord God, women comes to be coheirs, and fellow labourers, in the Gospell. . . . So here was the womens meeting, and womens teachings, of one another, so that this is no new thing."[53] This was obviously what the Women's Meeting in Yorkshire did: "We have cause to bless the Lord which hath called us weak ones into a spiritual exercise," the women wrote to the London Yearly Meeting, "and we find it our duty in our several meetings to exhort one another to be faithful and to be steadfast."[54] The importance of access to spiritual direction at the Women's Meeting is evident in Elizabeth Webb's advice to recently converted Friends: "Therefore I exhort and counsel you as one that have found favor with God that you meet often together and wait for the counsel of the spirit of truth."[55]

For those women who did attend meetings regularly, the spiritual directors became an indispensable source of spiritual consolation. We can almost feel the weight Joan Vokins placed on these meetings: "We shall not forget our times and hours to wait for the seasons of the Lord, for they are so sweet to the thirsty soul that it cannot be satisfied without them, and therefore many times thinks it long ere the meeting day come, that it might be replenished."[56] Spiritual direction was accessible to all who attended the meetings, even to those not fully expecting it or accepting it. Alice Hayes recounts in her spiritual autobiography that as a young girl she was invited by neighbors to attend a meeting and was exposed to Quaker spiritual guidance quite unexpectedly and in an overwhelming manner: "When I came to the meeting, it had made a great impression upon my mind, beholding the solidity of the people, and the weighty frame of spirit they were under, occasioned many deep thoughts to pass through my heart," Alice confessed. When "a woman stood up and spoke, whose testimony affected my heart," Alice "could not refrain from weeping."[57] Mary Proude Penington relates a similar experience of hesitation being overcome by the spiritual force of the women at these meetings. "I received strength to attend the meetings of these despised people which I never intended to meddle with, but found truly of the Lord, and my heart owned them. I longed to be one of them, and minded not the cost or pain; but judged it would be well worth my utmost cost and pain to witness such a change as I saw in them—such power over their corruptions." It took some time for her to respond to their example and guidance "before I came thither," but when she did, "Oh! the joy that filled my soul in the first meeting ever held in our house at Chalfont. To this day I have a fresh remembrance of it."[58]

The autobiography of nineteenth-century Elizabeth Comstock allows us to see how spiritual direction remained central to Quaker women's meetings even with the passing of generations. When her mother brought her at age nine to a meeting in London, "there for the first time, I had the privilege of listening to that eminent servant of the Lord, Elizabeth Fry. I shall never forget the impression she made upon my young mind by her sweet voice, beautiful face, and her earnest pleading." The desire to follow her exhortations to spiritual perfection through charitable works was firmly implanted in young Elizabeth when "in the solemn silence that followed, after she took her seat, my childish heart was left in the prayer that I might grow as good as she was, and work in the same way."[59] From Elizabeth Fry's own writing we get a sense of what the essence of her guidance was. "No person will deny the importance attached to the character and conduct of a woman, in all her domestic and social relations, when she is filling the station of a daughter, a wife, a mother, or a mistress of a family. But it is a dangerous error to suppose that the duties of females end here," she begins. Women are also spiritual people and as such have spiritual obligations to help others in any way they can. Elizabeth, however, adds a note in her direction to the women: "In endeavouring to direct the attention of the female part of society to such objects of Christian charity as they are most calculated to benefit, I may now observe that no persons appear to me to possess so strong a claim on their compassion, and on their pious exertion, as the helpless, the ignorant, the afflicted, or the depraved, of their own sex."[60] Elizabeth's concentration on getting women to help women is wholly consistent with the original focus of the Women's Meeting, for, as William Penn said two centuries before Elizabeth, these meetings were occasions "when by themselves [women] exercise their gifts of wisdom and understanding in a discreet care for their own sex."[61]

The eighteenth century saw the continued proliferation of sectarian groups, and a few even promoted the belief in a female messiah. Mother Ann Lee was promoted as a manifestation of the Second Appearance of Christ by her followers after her death, although she personally never claimed to be such. She did, however, establish a firm theological foundation for women's ministry in her United Society of Believers in Christ's Second Appearing (known as the Shakers), including women spiritual directors. "In searching the records of scripture we find that, on many extraordinary occasions, in past ages, there were females, as well as males, raised up and qualified to do the will of God and to accomplish his work," wrote Calvin Green and Seth Wells after Mother Ann Lee's death in their definitive and highly influential *A Summary View of the Millenial Church.*[62] "They were originally designed to have a correspondent share in teaching and guiding the human race, in directing the destinies of nations and governing the world, which is composed of females as well as males, who certainly stand in need of instruction, direction, and government."[63] The

Shaker document, *A Guide and Wall of Protection*, declares that all "souls set out in the way of God" must always be "in obedience to the leading and teaching of whomsoever it may please our Heavenly Parents to appoint— and when they counsel us, to do our duty, according to their counsel, we ought to fulfill it" for all "His witnesses" are "appointed to counsel us in love and mercy and meekness."[64]

The powers of Mother Ann Lee's spiritual direction were legendary. Hannah Cogswell tells us, "I have myself been an eye and ear witness" of Mother Ann Lee's ability "to search the hearts of those who came to see her" and to guide them "by the searching power of truth."[65] Generations later visionary Shaker Elder Rebecca Jackson reported still receiving counsel from Mother Ann Lee through other elder spiritual directors: "I received counsel from Eldress Pauline, as a preparation and endowment for a greater work. In speaking to me, Eldress Pauline said, 'These are not my words—they are right from Mother.'" Rebecca already knew this, "for this has been Her counsel to me for hours, days—yea, and months. I know they are Mother's words," Rebecca added, and "I will obey you in all."[66]

Among these eighteenth-century groups were the Methodists, and it is with them that we see spiritual direction the most developed and women spiritual directors most encouraged outside the Roman tradition. Like many other reform sects, Methodism began in the personal spiritual experience of its founder, in this instance that of John Wesley. It is only logical, then, to look to his chief spiritual director for the origin of his "method." That director is undoubtedly his mother, Susanna Wesley. A Nonconformist before she married an Anglican minister, Susanna had a strikingly independent theological mind. In her earliest surviving letters, Susanna relates first to friend Lady Yarborough and then to a minister who offered to lend her "direction in this particular" the details of a quarrel between herself and her husband. "Since I'm willing to let him quietly enjoy his opinions,"[67] Susanna could not understand why "my Master will not be persuaded he has no power over the conscience of his Wife." Susanna quite openly can see no "reason I have to ask either God Almighty's or his pardon for acting according to the best knowledge I have of things of that nature."[68]

This principle was the starting point of what she considered to be her chief mission in life, the spiritual direction of her children. It was a task she found daunting, for she well knew "what exemplary virtues are required in those who are to guide others in their way to glory!"[69] Susanna devoted "above twenty years of the prime of life in hope to save the souls"[70] of her children "by endeavouring to instil into (their) minds those principles of knowledge and virtue."[71] She made these, knowledge and virtue, the two prongs of her method of spiritual direction. (Many scholars posit that Susanna's method was the original and ideal method adopted by son John in his organization of the Methodists.[72]) In her journal Susanna clarified and strengthened the justification for her method. "Knowledge is indeed an admirable thing, as it is the foundation or basis of wisdom," which in turn "draws conclusions in order to practice them in particular cases." Thus virtuous behavior "is the good use you must make of your knowledge," for your interior life must "keep pace with your knoweledge."[73] Just how seriously Susanna took her task as spiritual director is also seen in a journal entry: "If any one soul among your children or servants should perish for want of your example or instruction, your own would be in danger of eternal damnation."[74]

Susanna strongly believed that the acquisition of knowledge and the practice of virtue demanded a very structured life. "Experience teaches you that 'tis absolutely necessary to spend a considerable time in spiritual exercises, and therefore be careful to get and improve all opportunities of retirement and recollection,"[75] she wrote in her journal. It should come as no surprise, then, to hear her advise her children even as adults that a good spiritual life was dependent on structured time management. "Let us converse as beings whose existence on earth is of short continuance and yet have a work of great, I should say, of the greatest, importance, to finish," she counseled son Charles. "This consideration will readily suggest to your good sense that we ought carefully to improve our time. And in order to do it effectively, I must earnestly conjure you to set apart two hours every day for private devotion; one in the morning, the other in the evening," she continued, but "it is not for me to fix the particular hours; they must be determined by yourself, who best know the method of your studies and what time you are least engaged." After all, "what wise man" would possibly "neglect such an excellent means of advancing spiritual life?"[76] She directed son Samuel to "examine well your heart and observe its inclination" to make sure his dedication to the spiritual life is total. "Let me tell you 'tis not a fit of devotion now and then speaks a man a Christian but 'tis a mind universally and generally disposed to all the duties of Christianity in their proper times" who achieves the goal of perfection.[77] Susanna advised son John even more specifically. "I see nothing in the disposition of your time but what I approve, unless it be that you do not assign enough of it to meditation," she critiqued, "which is (I conceive) incomparably the best means to spiritualize our affections, confirm our judgments, and add strength to our pious resolutions by any exercise whatever." It is apparent that already by 1734 Susanna's spiritual direction was sought not only by John but also by his friends, for in the same letter it is clear they all considered her as a spiritual director whose guidance they followed. "If you and your few pious companions have devoted two hours in the evening to religious reading or conference, there can be no dispute but that you ought to spend the whole time in such exercises," she answered to a previous inquiry of John's, "but if your evenings be not strictly devoted, I see no harm in talking sometimes of your secular affairs."[78] She cautioned John not to "rigorously impose any observances on others,"[79] just as she had individualized the spiritual supervision of her children.[80]

That Susanna took the pursuit of knowledge so seriously is no more evident than in a rather uncompromising early letter to Samuel in which she proclaimed that "if I thought you would not make good use of instruction and be the better for reproof, I would never write or speak a word to you more while I live."[81] Scripture was the beginning of knowledge, and, therefore, Samuel must "read and study constantly" and turn to it "in all cases that occur where you want direction."[82] While the children were growing up, at five o'clock "the oldest took the youngest that could speak, the second the next" and "read the Psalms for the day and a chapter in the New Testament; as in the morning they were directed to read the Psalms and a chapter in the Old."[83]

Susanna also encouraged the reading of devotional literature, but here she remained an active overseer. "*The Life of God in the Soul of Man* is an excellent good book," Susanna instructed John. "There's many good things in Castaniza more, more in Baxter, yet are neither without faults, which I overlook for the sake of their virtues"[84]; in another letter she instructs John on how to assess the merits of Thomas à Kempis's *Imitation of Christ*.[85] There are, however, some "studies [which] tend more to confound than inform the understanding, and young people had better let them alone," Susanna told John. "I can't recollect what book I recommended to you," she admitted to John another time, "but I highly approve your care to search into the grounds and reasons of our most holy religion, which you may do if your intention be pure, and yet retain the integrity of faith."[86] One must always remember the desired goal of knowledge, though, whether it come from Scripture, devotional books, or natural reason. "Suffer now a word of advice," Susanna cautioned John. "However curious you may be in searching into the nature or distinguishing the properties of the passions or virtues (of humankind) for your own private satisfaction, be very cautious of giving nice definitions in public assemblies, for it does not answer the true end of preaching, which is to mind men's lives, not to fill their heads with unprofitable speculation."[87]

We find in Susanna's spiritual direction an element common to the Roman tradition, that of the analysis of the spiritual life. If she personally engaged in such analysis we have no documentation of it, but her letters do reveal an awareness of such an approach. "The different degrees of virtue and piety are different states of soul, which must be passed through gradually; and he that cavils at a practical advice plainly shows that he has not gone through those states which were to have been passed before he could apprehend the goodness of the given direction," Susanna told John.[88] She seemed to warn him away from being too introspective about the stage of his spiritual life, because too often such identification is done at a superficial level and by physical manifestations: "Therefore, you must not judge of your interior state by your not feeling great fervours of spirit and extraordinary agitation, as plentiful weeping, etc., but rather by the firm adherence of your will to God."[89]

Susanna placed great emphasis on meditation and was quite emphatic that even she "keep the mind in a temper for recollection and often in the day call it in from outward objects, lest it wander into forbidden paths."[90] Before beginning meditative prayer one must "take at least a quarter of an hour to recollect and compose the thoughts before your immediate approaches to the great God; one prepares for secular rulers so how much more should you take care to have your mind in order, when you take upon yourself the honour to speak to the sovereign Lord of the universe?"[91] She knows that meditation is but the first stage toward God and often leaves one "yet unsatisfied because you do not enjoy enough of God," and frustrated at apprehending "yourself at too great a distance from him by faith and love."[92] Susanna seemed to have no experiential knowledge of prayer life beyond that and evinces no interest in directing anyone to the mystical life. She is intent, on the other hand, upon guiding others to meditation through self-examination. "I heartily wish you would now enter upon a serious examination of yourself,"[93] she advised John as he began his studies for ordination, while Samuel she directed to "examine well your heart" and remember that "the mind of a Christian should be always disposed to hear the still small voice of God's Holy Spirit, which will direct him what and how to act in all the occurances of life."[94] She wanted all her directees to "make an examination of your conscience at least three times a day and omit no opportunity of retirement from the world."[95]

Given the influence Susanna exercised over John and the years she was his spiritual director, it would be surprising if spiritual direction were not of prime importance within Methodism. John's correspondence indicates he frequently turned to women for spiritual guidance, particularly to Selena, Countess of Huntingdon. He apparently trusted her direction so much that he submitted his spiritual journal to her for approval before publication. (Twice she wrote of her approval, suggesting only that he diplomatically dedicate "an address to the bishops and clergy."[96]) He also followed her advice to compile "a chaste collection of English poems" as a means of teaching "the distinction between virtue and vice." When he did so he wrote to tell her "I inscribe these poems to you, not only because you was [*sic*] the occasion of their thus appearing in the world, but also because it may be an inducement to many to read them."[97]

Sarah Crosby employed one of John's own favorite methods of spiritual direction in her direction of him: the personal catechism. "I know the Lord Jesus loves you, and that you are a chosen Vessel," she began, but "do you love the Lord Jesus more than any person or thing? Do you find more happiness in thinking or speaking of him, than in thinking or speaking of any creature? Does your soul delight in him? This is what my soul desires for you."[98] Sarah Ryan used the same approach with John. "How I long for you to be holy, in spirit, soul, and body! Has God fixt a resolution in your soul, to grasp the glorious prize? Do not you depend too much upon any creature?" she prodded. "Dear Sir, use much private Prayer, and much good will come out of all this."[99] Historian Earl Kent Brown reports that in a study of the sixty-three early Methodist women there are extant sources for, sixteen were known to practice spiritual counseling.[100]

In the same study Brown also reports that forty-five of the sixty-three women were leaders of Methodist classes and bands. Although the idea of small-group communities is certainly not original with Methodism, no sect from this era used them to more advantage. Bands were small same-sex voluntary groups formed specifically for mutual spiritual direction and encouragement, while classes were mandatory and sometimes mixed, but still were concerned with spiritual growth.[101] From the beginning women held leadership positions in both. Sarah Crosby was such an admired leader of a Leeds class that over two hundred people attended where she regularly told them "part of what the Lord had done for myself, persuading them to flee from all sin."[102] At age seventy Sarah was still directing "two classes of about thirty persons each, every week, and two or three bands."[103] Lady Maxwell was a well-respected leader of a class composed of preachers and their wives.[104] John Wesley admired Hester Roger's ability to direct classes so much that he encouraged her "to watch over the new-born babes" with "much light"; she was eulogized as a woman with many talents, "particularly as a leader of classes and bands."[105] In a letter to John, Hester informed him that while in Nantwuch the people "flocked around me with eagerness" and that through her guidance " a poor backslider was restored and all present were filled with humble love and joy. I left five or six earnestly crying for a clean heart."[106] Hester tried to engage one backslider in regular correspondence, because she perceived the need for constant direction: "You see how freely I write, as if I had known you seven years. I hope you will follow my example in this, and let me know the particulars of your spiritual state."[107] Mary Fletcher began a class in Leytonshire before the Methodist had even established a society there.[108] Lady Huntingdon went so far as to override John Wesley's instruction to Thomas Maxfield that he lead a London class in prayer; she advised Maxfield to preach to the group. Indeed, she wrote John that "I made him expound"[109] and used his success as an example in her spiritual advice to

John that he allow lay preaching in the groups.[110] George Whitefield, who frequently turned to Lady Huntingdon for spiritual advice, hints that many others also did so when he describes her as "acting the part of a mother of Israel" more and more. "For a day or two she has had five clergymen under her roof, which makes her Ladyship look like a good archbishop."[111]

Sects with women directors were not peculiar to England; they also thrived in America. When George Whitefield and Gilbert Tennent spread the Great Awakening in Newport, Rhode Island, one convert was Sarah Wheaton. Soon after her initial exposure she was called upon to establish and direct a women's meeting that would meet weekly throughout the remainder of her life: "A number of young women, who were awakened to a concern for their souls, came to me, and desired my advice and assistance, and proposed to join in a society; provided I would take care of them. To which, I trust with a sense of my own unworthiness, I joyfully consented."[112] In a confessional letter to one Joseph Fish, Sarah tells him that "I know of no one in the town now that is against me. My dear Mrs. Cheseborough and Mrs. Grant Have both been to see me and thank'd me for persisting Stedily in the path of duty against their discouragements, ownd they were at first uneasy but now rejoicd and wish'd a blessing." Her direction was so appreciated that "every Intimate brother and friend intreats and charges me not to dismiss So Long as things rest as they are, telling me it would be the worst days work that Ever I did if I should, as God His Self Has thus Employd me." In fact, "other Masters and Mistresses frequently send me presents in token of gratitude" for the changes her spiritual direction has wrought in their community. She admitted the changes were drastic: "From unwillingness to Learn or know any thing good, they are now intent upon Learning to read etc. at Home and abroad; some that were unwilling to serve and saucy are become diligent and condecending; some that were guilty of drinking gaming Swearing Sabath breaking and uncleanness are at present reform'd; Several couple Latily Married who had Liv'd together without but could not bear to Live in the Sin any Longer."[113] Sarah sincerely believed "that this Gatherings at our House" under her spiritual direction in "no way tend to Separations rents or diversions but are rather a Sweet Sementing bond of union that Holds us together in this critical day."[114]

Chapter 9

The Modern Era

Within the Roman tradition women spiritual directors flourished during the eighteenth and nineteenth centuries. Whereas few Protestant women directors were self-conscious of the ministry they were engaging in, many Roman women were very aware. They openly identified themselves as spiritual directors and were most deliberate about their task. They relied heavily on the spirituality and methodology of past spiritual directors, particularly that of Teresa of Avila, for their models, but they always adopted past practices to their specific situation. It was within the religious orders that the women spiritual directors found a permanent home, and, because women religious orders were multiplying exponentially during these centuries, adapting direction to the needs of each order was challenging. New orders were founded almost annually, each with a unique task it defined as its own. Nursing, missionary work, teaching, contemplation, social work, work among slaves and Native Americans, the administration of orphanages and camps, catechetical work, sewing, clerical work, editorial work and translation, bookbinding, caring for the aged, hospices work: Each of these tasks had a religious order dedicated solely to its completion.[1] And because each order defined its apostolate differently, each consequently developed its own spirituality centered on that apostolate.

Likewise, because every order saw its spirituality as peculiar to itself, it turned inward to ensure that its particular spirituality would be transmitted intact to the next generation of members. Every religious order desired its own spiritual director familiar with their specific spirituality. They responded to this need by reinforcing the position already designated for this task, the novice mistress. A formalized novitiate under the close direction of a novice mistress whose job it was to guide the aspirants according to the order's spirituality became a universal institution of these new orders. Not only did the new orders revere the traditional position of spiritual director, then, but they even increased its importance by assigning it with the task of preserving the order's unique spirituality.

Confessor-directors did not disappear from the scene, but their sphere of direction was limited in many ways. A Franciscan confessor-director might well guide women religious through the difficult stages of prayer, but he could hardly be expected to know how to direct a member of the Marist Missionary Sisters with the spirituality necessary to inform her work in maternity centers, an apostolate peculiar to those sisters. On the other hand, although the need of the religious orders to direct novices in the group's spirituality gave enhanced prominence to novice mistresses, they could and most times did also engage in the direction of prayer life. This was especially true of the original founder of an order, who defined the order's spirituality at its origin and directed the first members toward its end.

We can see this in numerous sources left by women religious, especially those from the prolific nineteenth century. The correspondence of Madeleine Sophie Barat, founder of the Society of the Sacred Heart of Jesus, provides us with an excellent opportunity to see how spiritual direction thrived within new religious orders. Madeleine Sophie was a consummate administrator and organizer, but, as she states in one of her earliest surviving letters, these activities were secondary to the direction of souls. "I am longing to hear news about you, with details," she wrote one of her nuns, Mother Philippine Duchesne. "You know what interests me above all, the progress of your soul. Speak to me more about that than about the business of the house, although you know that I am not indifferent to that."[2] Madeleine sincerely believed that spiritual direction was one of the most significant and rewarding tasks a person could undertake. "Happy the simple and straightforward soul, free from self interest, who gives to others the light that she has received without keeping anything for herself alone, any more than the channel that gives water from a spring," Madeleine proclaimed to another superior, Mother de Curzon. "The Spirit of Jesus, who always dwells in an interior soul united to the Divine Heart, will make us know what to say, decide, advise. We are then docile instruments that receive and hand on."[3]

Because of the profundity of the apostolate, Madeleine made spiritual direction one of the chief purposes of the Society. To a struggling novice named Emile, she explained how intertwined spiritual direction was to personal happiness. "I think that you do not sufficiently realize the sublimity of your vocation, for there are no labors, however hard they may be, that you should not be ready to carry on to reach your end which is, as you know, to save souls. Those entrusted to you must be constantly the object of your care. To make them truly pious you must burn with love for your God, and take from your heart all that opposes that love: pride, sensitiveness, laziness."[4] When Octavie Berthold was named mistress of novices, Madeleine bolstered her spirits in the face of this new responsibility. She repeated the message she gave to Emile and reminded Octavie that "if you are still the mistress of novices, what an important and holy employment! And how holy you yourself must be in order to inspire those souls with the spirit of the Society; more by your example than by your words."[5]

According to Madeleine, the women must always remember an essential reality in spiritual direction: Jesus is the ultimate spiritual director. "What a sublime vocation, to save souls!" Madeleine explained to Elizabeth Galitzin, a prospective member of the Society. "But the instruments must be worthy of their heavenly work, and how much you still have to acquire! Thank God you are in the arena, and Jesus Himself will be your guide and your helper."[6] The women must do what they can to direct people here and now, but the human is frustrated in the attempt to guide another, because "Jesus alone understands our hearts."[7] When advising Josephine Goetz on how to best fulfill the ministry of spiritual direction among novices, she summarized it thus.

> The less there is of your own action, the more will the good Master supply with His own, and despite the number (that you have to guide) and of the shortcomings of each one, everything will move ahead with ease. For when there is question of directing souls the action of God is needed. All the elements obey man, so to speak, but the Creator and Redeemer has kept the realm of souls for His own. He wants to make use of us, certainly, but only as instruments and not as movers. Let us allow Him to act. Let us be no more than a gardener who cultivates the soil; he turns it over and pulls out the weeds; but once the seed has been sown He has no more to do than to water it and to drive away the insects and other enemies of the plant.[8]

To acknowledge Jesus as the ultimate spiritual director did not lead to less emphasis on human spiritual direction, particularly that given in the novitiate. Here the order's spirituality was presented and potential members were guided accordingly. "If regularity and fervor reign in your house, be sure that Jesus will bless it and will send you novices," Madeleine wrote to one novice mistress. "But if you slacken in virtue, if each one prefers her own interests before those of Jesus, all will relax and wither up."[9]

Madeleine Sophie's own direction of women was perhaps the best model for her Society's novice mistresses to imitate. She knew the importance of personal sanctity and example in the direction of others. "Saints make saints," she once commented. Since "we only form what we are ourselves are," we must first "break down all the obstacles that are opposed to our sanctification,"[10] for "how precious is the time that is left to us, and how important it is for us of the older generation to become models while we are teaching others."[11] There was no better way to become a model than to imitate the models of the past, Madeleine argued, for that was "the reason for His love and His liberality with regard to the Teresas, the Gertrudes, the Catherines of Genoa and Siena, and so many others that you know." Just as "this great God who so loves to communicate Himself" revealed the purpose of life through these women, so too will he "still do the same if He found hearts as well disposed."[12]

In fact, all members of her society must remember "that Our Lord wants you to become a great saint" whose example others will follow, and because of this "He has brought you into our little family in order to give you the means to this."[13] Madeleine constantly called on past women saints in her direction of others, Teresa of Avila being perhaps the most frequently referred to. "I love this saint, her spirit of Jesus, her love, for Him her interior life which we should imitate," she wrote in gratitude when given a relic of Teresa. "I love to find among my daughters souls who have thought for a time of joining Carmel, because they have the foundation of interior life, which, joined with the apostolic life, makes them excellent religious."[14] Like Teresa's reform Carmel (Carmel is a generic name for a Carmelite monastery), "our society which is made for both these works—the cultivation of prayer and the salvation of souls—must also draw its strength from the help of contemplation. It must do this the more essentially as the century in which we live is more difficult that in the century of St. Teresa."[15] Thus, in the aftermath of the antireligious strains of the French Revolution, Madeleine directed her novices to imitate Teresa, who "from the depths of her solitary cell took in the entire world in her desire to win it for Jesus Christ. Our hearts must be no less vast than her own."[16] Teresa was able to accomplish such greatness "from prayer, from her union with God." Just as Teresa "placed prayer as the basis of her Institute," Madeleine directed the novices at Poitiers to do likewise. "For us who are called to the same perfection, that of sanctifying ourselves, do you think that we can reach this double end with a prayer and union with God in all our actions?"[17]

The examples of many other saints, male and female, were utilized by Madeleine in her spiritual direction. She chose Jane de Chantal as the model for "the virtues needed by a superior who is establishing a house." Particularly Madeleine emphasized "what gentleness, what patience" Jane displayed "at every moment."[18] Madeleine modeled the rule for her Society of the Sacred Heart on "those of St. Ignatius," a fact everyone knew "since it is impossible not to see that they come from the same source."[19] She received novices without doweries "as once St. Francis Borgia received the little Stanislaus Kostka,"[20] and she urged her Sisters to embrace willingly all trials and tribulations in imitation of "St. Catherine of Siena [who] chose the crown of thorns" and "found delight in the pain that she suffered."[21] Given that the Society's constitution of 1815 states that "the object of this Society is, therefore, to glorify the Sacred Heart of Jesus, by labouring for the salvation and perfection of its members" and "the sanctification of others,"[22] it is not surprising that Madeleine had a keen devotion for those women who promoted the devotion in the past. Margaret Mary Alacoque, the modern disseminator of Sacred Heart devotion,[23] was one of Madeleine's favorite models, as was Mary Magdalene, "my patroness."[24] Gertrude who "says wonderful things,"[25] about the Sacred Heart, was often held up by Madeleine as a model in her spiritual direction of novices. She send Adrienne Michel, her directee, one of Gertrude's books, saying, "you will like this reading; try to profit by it. I hope it may convince you of the necessity of purifying earthly affections if we wish to attain—I do not say to singular favors, for God does not grant these to every one—but to union with Him who is the end for which we were made."[26]

While Madeleine used all these saints as models in various situations, when it came to direction in prayer, Teresa of Avila was Madeleine's chief guide. We see how completely Madeleine adopted the spirituality of Teresa in her direction of Adrienne Michel. First, Madeleine reminded Adrienne of the importance of spiritual direction. After thus setting the stage, Madeleine then introduced Teresa's distinctions on the stages of prayer: "It is a matter of mortifying ourselves continually until we have acquired such a mastery over our exterior senses that the soul is in control. Then she is approaching the state of entire purification which leads to union with God." Madeleine also recalled Teresa's conviction that "many holy souls have attained to union with God, without having been enriched with these gifts. I am speaking of the peace tasted by the heart that refuses nothing to the Lord." Because the ways of the spiritual life are complex, Madeleine ended her letter with a cautionary note, telling Adrienne not to pursue these matters without direction and a promise that "when I see you again we shall read St. Teresa together."[27] Adrienne, however, read Teresa without direction, and, because she did not submit to Madeleine's guidance, Adrienne confused Teresa's work. This prompted a rather severe reprimand from Madeleine: "What are you thinking of, little girl? You ought to be scolded for such a thing! You are only at the ABC of the spiritual life, in theory and in practice, and you want to be like the doctors and fathers of the Church who only studied these works after having practiced the science of the saints for a long time! Keep to Rodriquez and Dupont, and those authors whom your superior has given you. The simplest is the best for you, who always try to fly too high on your little feeble wings. Simplicity! Simplicity!"[28]

As good a model as Teresa was to imitate, the uninitiated often needed a personal guide to point out the differences between the great saint and the beginner. "Ah, I understand how you can desirer [*sic*] to die, to enjoy with division and without end the embraces of God," Madeleine sympathetically advised Adrienne. "St. Teresa wished for this so vehemently that the strength of her words, energetic as they were, failed to express for her the devouring desire of being united to her Spouse. Should we desire it as she did? Alas, that does not depend on us; we would have to have her love, and above all to have worked as she did. We would have to have her great heart, her noble and generous soul, and we are so weak, so small, so strunken."[29] While she tried to help Adrienne to realize the differences between Teresa and herself, Madeleine still prompted her to imitate Teresa when appropriate. "Oh, if you possessed this passion," the selfless love of God that Teresa manifested, "how your soul would grow large; how it would rise above the low regions of the sensible and the natural," Madeleine ruminated. "Alas, my daughter, we have lived long enough in that region; is it not time to rise higher?"[30]

From the very beginning of her ministry in spiritual direction Madeleine was intensely aware of "how few understand true spirituality," no less the masterpieces of spiritual literature. She was most cautious, therefore, in her direction of spiritual directors. "Be careful not to go ahead of grace in the case of Clemence and of others," she wrote to one of the Society's directors. "With regard to the prayer of quiet, the Holy Ghost must be the one to give it, and in the beginning it is not always continuous." The director must step back and simply "prepare such hearts by helping them to get rid of their passions. For, believe me, my daughter, it is in vain that one enters into the way that makes contemplatives, and one will not reach this end, unless the heart has been purified by the annihilation of self. Take care of that before leading souls too quickly into the way of quiet. I give you this advice in passing, because for some years now a few souls who, no doubt, did not understand you came very near to illusion."[31]

Madeleine's correspondence also reveals another essential element in spiritual direction, that of spiritual criticism.[32] She had unwavering confidence in her own ability to direct others, and so she rarely hesitated to be harsh or condemnatory when needed. She warned Elizabeth Galitzin of her approach in spiritual direction, but also promised results. "Do not be troubled about the difficulties and the distastes that you feel sometimes in prayer," Madeleine wrote to Elizabeth as she awaited entry to the group. "I don't want to excuse you altogether, and you should sometimes be punished—but from so far! We must wait, and I promise you, my daughter, that when it shall please God to put you under my guidance I shall spare nothing in you; for I am very eager that you should make great progress in the practice of solid and religious virtues, and thus respond to the immense love that Jesus Christ has for you."[33]

We can see how faithful Madeleine was to her promise to spare Elizabeth nothing in a letter of admonition she wrote years later when the latter was a mother superior herself. First, Madeleine chastised Elizabeth for "writing by every mail to ask the same thing over again" and told her she must "pay the postage yourself; when you see what your repetitions cost you will think twice about them; at least, I hope so."[34] Then Madeleine addressed Elizabeth's request to fulfill her apostolate elsewhere: "I answer with a point-blank 'no.' Really, my daughter, how foolish. Get the idea out of your head and leave us in peace. Moreover, what would you do there? You are not prepared to guide either religious or children."[35] When directing Mother Eugenie Aude by mail, sometimes the entire letter was filled with reproofs. In one letter Madeleine warned Eugenie to "take care of your health and do not be so eager to die," for these attitudes "show more love of self than of Jesus Christ." Eugenie must instead "work to gain souls for Him" and to spread the Society: "So, don't be imprudent." That was only one of her problems, though, so Madeleine continued: "Neither am I satisfied with your false humility and with your insistence that you be relieved for your office of superior. You are afraid of trouble and of responsibility." After revealing the error, Madeleine then guided Eugenie in the path back to perfection. She must realize that by taking on "a painful cross," she would realize her apostolate is a gift and "thank Him for having enough respect for you to entrust to you souls who are dear to Him."[36]

Often her criticism was less biting, although just as essential. "Remember the little piece of advice that I gave you six months ago: you look too much at yourself, at your miseries and your unworthiness; you are afraid of not being humble enough, and when you are brought to the fore you feel obliged to hide under the ground and to reproach yourself without measure," she summarized to Pauline de Limminghi. "Believe me, my daughter; don't make such a fuss. Turn all that to Our Lord by a calm and simple look, and let people talk! Don't think about it, forget yourself, forget even your own perfection, so to speak in order to see only Jesus and His glory." In case Pauline doubted her direction, Madeleine supported it with Scripture: "Your fault is found in Chapter V of the Canticle where the bride fears to rise to open to the Bridegroom who knocks. Read it."[37] Thus we can see operating in the spiritual direction of Madeleine Sophie Barat one of the chief characteristics of spiritual direction, adaptability to specific situations. When gentleness, simplicity, and love were called for, they were used to direct, but if harsh discipline was needed, it would be employed appropriately. Madeleine encapsulated this approach in remarks concerning the Society's direction of nineteenth-century French laity.

> What tears my heart every day is the spectacle of so many souls who are lost by their own will and in spite of the help given to them by religious—your persons nourished by and brought up in piety whom pleasure draws away! In

one year only, what ravages in these young hearts! Alas, it is no longer the love of God that we must teach them; that does not touch them. We must bring remorse into their hearts, and threaten them with eternal punishments. In a pressing danger perhaps they will remember this; it is almost the only hope. Oh, how much more consoling it would be, as you say, to make Jesus Christ known to fresh hearts.[38]

What makes Madeleine Sophie Barat's letters of spiritual direction so significant is precisely that they are not exceptional. She was just one of many thousands of women spiritual directors who functioned in the new religious orders of the period, quietly repeating the principles established by past masters, sometimes with innovations, but most times simply by adapting the long tradition of spiritual tradition to their own situation. Hundreds of orders were founded in the nineteenth century with innumerable novitiates. Women spiritual directors abounded and flourished in each one of these houses. Only a very minute proportion left sources behind, but there is enough for us to realize how commonplace the woman spiritual director had become by the nineteenth century. Many of them never received notice outside their convent walls, but within them numerous cults around these women grew. More than a few of these cults were pursued to the canonization process; many are still being pursued. The vast majority of them, nevertheless, had their memory and their contributions in spiritual direction preserved only by historians of their orders.

The few works that were published and popularized contain elements of spiritual direction similar to those found in Madeleine Sophie Barat's letters and conferences. For example, Frances Mary Teresa Ball, founder of the Institute of the Blessed Virgin Mary in Ireland, was as dependent on spiritual models in her direction of members as Madeleine Sophie Barat was. Frances's admiration of Teresa of Avila led her to promote a eucharistic devotion among the members of her Institute. Adhering to Teresa's conclusion that "the most important moments of our life are those after Holy Communion,"[39] Frances never missed an opportunity to advise a directee of its value. "Do not omit Holy Communion"[40] even when in spiritual difficulty or temptation, Frances tells one nun, because she tells another directee, "Holy Communion is no less beneficial though our minds may be beset with temptations. It is true the more we desire this heavenly food, the greater grace we shall obtain." Accordingly, "I would not omit one Communion for which I had leave, for we are accountable for the graces we should receive if we approached."[41] To another woman Frances explained her love of solitude thus: "St. Teresa likes hermitages, so do I."[42]

Like Madeleine, Frances did not limit her models to Teresa or to women. She proposed that Catherine of Genoa be the model for the mixed life of the

Institute because "Catherine of Genoa, though actively employed, ever thought of God."[43] Ignatius of Loyola was another of Frances's models. She directed Mother Mary Teresa to follow the example of Ignatius when guiding and ruling others. "What have we to fear when God appoints us to govern? Is He not powerful enough to assist us, sufficiently wise to guide us?" Frances queried. "And St. Ignatius tells us to do the work as if success depended on our own efforts; but to be certain that God alone can crown our operations for His greater glory."[44] In another incident Frances turned to Ignatius' directives as the basis of some of her own decisions: "With St. Ignatius, I think if there is a pure intention, the usual times for prayer suffice. I would not wish for the singularity of two Retreats of eight days in the year."[45] Again like Madeleine Sophie Barat, Frances placed great emphasize on spiritual direction and told her directees to "ask as often as you please any question that will contribute to your peace."[46] She accepted her duty to direct others, well aware of its responsibilities. "You did well to unfold to me your ideas, apprehensions, and uneasiness. The will of God is that one should guide and assist the other to heaven, and the miseries of life lessen in proportion as we unbosom our mind to one appointed by God himself to be our leader in the dreary path of this exile from our true country."[47]

The letters of Henriette le Forestier d'Osseville provide yet another example of the kind of spiritual direction commonly practiced within religious orders of the day. It was not particularly creative or original, but it did meet the basic goal of direction; it guided directees to happiness by giving all their thoughts and actions meaning. In editing some of Henriette's letters John George Macleod admits he chose to edit them "not as containing any incidents of interest or spiritual instructions of a strongly marked or original character,"[48] but rather because they were typical of the kind found in most orders' archives. We find in Henriette's direction some of Madeleine's nononsense approach to direction. "Yes, more than ever I forbid you to be discouraged. A missionary alone in China, without a guide, or spiritual help, or advice, or a mother to love her, has more to suffer than my little English missionary," Henriette shamed one directee working in England who was complaining about the lack of spiritual success. Henriette wrote in the same vein to another discouraged member. If she was tempted to believe herself overwhelmed with burdens, then she must "go to the Garden of Olives, to Mount Calvary, to the Cross," and see Jesus suffer, "And me too, in a little corner, showing Him to you, and asking you if you have suffered that sweat of Blood, or the agony afterwards from which He rises exhausted to go and suffer the torments of His passion. Never may there be any discouragement or melancholy," Henriette ended. "Oh, what cowardice!"[49] To yet another member resistant to spiritual direction Henriette took an even harsher line. "You must do with the children that God has confided to you. He counts on you, and so do I. Are you going to disappoint us and become a spiritless woman, without energy and without faith? But you have faith, and I appeal to it, and to your heart," Henriette implored her, "to rise immediately on reading this letter. My child, YOU MUST. *Obey your Mother.* Rise, go to the chapel, and pass from it another person—a valiant woman, a daughter according to my heart."[50]

Of course, not all her direction demanded such stringent approaches. Most of it, again as we saw in the lengthy discussion of Madeleine's direction, dealt with ordinary matters. Henriette had great faith in her ability to deal with such direction through correspondence, and she encouraged her members to avail themselves to this means. "I like the letters that you write with such simplicity and confidence. I read your soul in them, and that is what I want," she commented to one directee. "Write in this way, but do not reason so much on your state, nor take too close account of it," because that was her job as director. She continued: "Go on with your eyes shut, by the path which is shown you, and I promise you I will not let go of you."[51]

It is fitting that the final discussion of a woman spiritual director offered in this survey is not typical but original and highly influential. It was because of her exceptionality that Pope Pius X, a mere seventeen years after her death, called Therese Martin of Lisieux "the greatest Saint of modern times."[52] She lived an extremely obscure, brief, and uneventful life in an extremely narrow, private, and isolated circle of acquaintances. Yet, amazingly, within fifty years of her death she had churches dedicated to her on every continent, was named the patron of France and the universal patron of missionaries in the Roman church, and had her autobiography published in innumerable languages and editions worldwide. She is without question one of the most popular saints in the Roman church. Because of an unsophisticated presentation heavily laced with sentimental, childlike imagery, her writings have been the subject of much scholarly debate throughout this century. Many claim her understanding of ultimate realities is expressed in such an affectatious manner that it distorts the profundity of those realities. Other critics claim her romanticized, emotional presentation of humanity's search for happiness verges on being antirational.[53] Therese's supporters argue that her excessive sensibilities to the emotional aspects of human nature are precisely the reason why her direction was so accessible to the masses. Supporters also argue that when patiently examined, Therese's grasp of happiness, however overly sentimentalized in its expression, possesses extraordinary intellectual depth. Dorothy Day, socialist activist and founder of *The Catholic Worker,* offers her personal testimony to the latter argument. In her initial exposure to Therese's spiritual directives, Day was somewhat repulsed by Therese's religiosity. Curious about the saint's popularity, though, Day returned later to Therese, determined to discover the core of her spirituality. When she did read Therese a second time, she found the answer. Therese's directives exercised utmost influence over Day from that point on, an influence Day describes in detail in her book *Therese.*[54]

Critics and supporters alike agree on some pertinent facts. They agree that Therese's spiritual direction was unique in its ability to address such a vast and diverse audience. She was able to identify the universal elements necessary for happiness as accurately as any philosopher or psychologist and express that discovery in such a universal manner that it transcended the particularities of different cultures. She simplified her direction to the lowest common denominator, and in so doing, she made it accessible to more people than anyone had ever done before. Her goal was the common goal of humanity, that of happiness: "I have always wanted to be a saint." This desire alone, though, was not enough to make her important in the history of spiritual direction; it was her method that did that. Because she was keenly aware of the ordinariness of her endowments, and because she knew most people shared her ordinariness, Therese decided that the best way for her and those like her to seek happiness was through ordinary ways. She called her method the little way: "I want to seek out a means of going to heaven by a little way, a way that is very straight, very short, and totally new."[55] After years of personally testing her little way, a few months before her death Therese confided in her sister Pauline (Mother Agnes of Jesus in religious life) that she believed that "my mission is about to begin, my mission to make God loved as I love Him, to teach souls my little way."[56] To that end Therese asked Pauline to oversee the posthumous publication of her spiritual autobiography, the vehicle through which Therese directed others along this "totally new" "little way."

After living a very quiet family life for fifteen years, Therese entered the Carmelite convent at Lisieux. Some twenty women lived there, and Therese remained within that monastery until her death from tuberculosis nine years later. The convent itself was a family affair, for before her death four Martin sisters and a cousin were members. Therese tells us this was at times "the source of great suffering," given the regulations of convent life. "I didn't come to Carmel to live with my sisters," despite how it looked, "but to answer Jesus' call. Ah! I really felt in advance that this living with one's own sisters had to be the cause of continual suffering when one wishes to grant nothing to one's natural inclinations."[57] Therese told Pauline that "I did not feel, as formerly, free to say everything to you for there was the Rule to observe." After entering Carmel, "I was unable to confide in you" because now "I was in Carmel and no longer at *Les Buissonnets* [the family home] under the paternal roof."[58] Therese's remarks here direct our attention to the first fact to note concerning her and spiritual direction: Pauline was Therese's first and most influential spiritual director. When Zelie Guerin Martin, their mother, died, Therese was four years old, and Pauline became Therese's "little Mamma."[59] Therese's opening sentence to her spiritual autobiography is addressed to Pauline, that is, "to you, dear Mother, to you who are doubly my Mother, that I come to confide the story of my soul."[60] It was Pauline who made "the attractive little book" for Therese's First Communion preparation and "aided me in preparing my heart through a sustained and thorough method,"[61] and it was Pauline who counseled Therese "to commence living a new life" immediately after "one received one's First Communion."[62] Likewise, it was "Pauline, too, who received all my intimate confidences and cleared up all my doubts" by bringing "the most sublime mysteries down to my level of understanding and [was] able to give my soul the nourishment it needed."[63] After Pauline entered Carmel she continued her direction of Therese by writing "me a nice little letter each week and this filled my soul with deep thoughts and aided me in the practice of virtue."[64] When Therese followed Pauline's example and entered Carmel years later, Therese acknowledged her debt to Pauline's direction, telling her "without you, perhaps I'd not be in Carmel."[65] Therese responded to Pauline's direction by "open[ing] up to you my poor, little, wounded soul. To you who understood it so well, and to whom a word, a look were sufficient to explain everything! I surrendered myself completely."[66] In her last months Therese reflected upon Pauline's lifelong direction and said she hoped "[Pauline] will realize that you have been, and will always be, the angel charged to guide me and announce the Lord's mercies to me!"[67]

Indeed, male spiritual directors had little influence on Therese during the entire course of her life. In a letter to Pauline Therese makes a passing reference to the ordinary confessor at the Carmel, but he obviously exercised minimal influence on her spiritual life. She brings the attention right back to Jesus after mentioning him: "Tomorrow, I'll go and find M. Youf. He told me to give him a little review since I am in Carmel. Really pray in order that Jesus may leave me the peace which He has given me. I was happy to receive absolution on Saturday."[68] Another priest, Father Pichon, claimed he spiritually adopted Therese at the time of her First Communion.[69] Therese confirmed this in her autobiography, but it was apparently of little consequence. She wrote him after her First Communion "telling him that soon I would be a Carmelite and then he would be my director. (This is what happened four years later, since it was to him I opened my soul)."[70] After she entered the Carmel he did come to direct her, but the interview "was veiled in tears because I experienced much difficulty in confiding in him." Nevertheless, she made "a general confession, something I had never made before," after which he declared that she had "never committed a mortal sin."[71] As consoled as she was by his words, Therese immediately lets us know that in reality Father Pichon was not truly her spiritual director.[72] She considered

Jesus alone to be "'my Director.' I don't mean by this that I closed my soul to my Superiors; far from it, for I tried always to be an open book to them."[73] The problem was that "it was only with great effort that I was able to take direction" from any except "the Director of directors."[74] It was "'my Director'" who bore Therese's faults patiently, "for He doesn't like pointing everything out at once to souls,"[75] just as it was "'my Director'" who taught her lessons about poverty.[76] "O Jesus, my beloved who could express the tenderness and sweetness with which You are guiding my soul!"[77] Therese exclaimed in gratitude. The extent that she considered Jesus her personal spiritual director is seen in her remarks as she pondered her own desire to direct others: "Why do I desire to communicate Your secrets of Love, O Jesus, for was it not You alone who taught them to me . . . ?"[78]

This is not to say that Therese had a negative attitude toward human spiritual direction, but she did seem to have some reservations about the way it was practiced in her own milieu. "I do not hold in contempt beautiful thoughts which nourish the soul and unite it with God; but for a long time I have understood that we must not depend on them and even make perfection consist in receiving many spiritual lights," she criticized. When Therese was charged with the direction of postulants and novices, it was an opportunity for her to practice her own type of spiritual direction. In her autobiography Therese began a discussion of her role as spiritual director by confessing that "when I was given the office of entering into the sanctuary of souls, I saw immediately that the task was beyond my strength."[79] Once she "understood that it was impossible for me to do anything by myself, the task you [the superior] imposed upon me no longer appeared difficult. I felt that the only thing necessary was to unite myself more and more to Jesus."[80] Like so many women directors before her, Therese soon concluded that only when she allowed her words and actions to be used "as instruments to carry on His work in souls" was her direction successful. "From a distance it appears all roses to do good to souls, making them love God more and molding them according to one's personal views and ideas," she admitted. "At close range it is totally to the contrary, the roses disappear; one feels that to do good is as impossible without God's help as to make the sun shine at night. One feels it is absolutely necessary to forget one's likings, one's personal conceptions, and to guide souls along the road which Jesus has traced out for them without trying to make them walk one's own way."[81] Therese's second insight was basic common sense: "I learned very much when carrying out the mission you entrusted to me; above all I was forced to practice what I was teaching to others."[82] She never forgot the lesson.

While Therese maintained that all peoples are created spiritually equal, regardless of their vocation or status in life, and that "the infinite riches of [Jesus'] Heart will supply for all and remove all unequality,"[83] she also knew that spiritual equality did not mean that everyone was treated the same. She realized immediately upon beginning her role as spiritual director that "it is impossible to act with all in the same manner," because "all souls have very much the same struggles to fight, but they differ so much from each other in other aspects."[84] As many spiritual directors before her realized, Therese knew her spiritual direction must be flexible and individualized. "With certain souls, I feel I must make myself little" by revealing to her directees her own faults, so that "my little Sisters in their turn, admit their faults and rejoice because I understand them through experience. With others, on the contrary, I have seen that to do them any good I must be very firm and never go back on a decision once it is made."[85] Most important, a spiritual director must acknowledge that her "mission was to lead [directees] to God."[86]

It is, of course, the method Therese proposed to lead her directees to God that makes her so significant in the history of direction.[87] When Marie (Sister Marie of the Sacred Heart) sought her natural sister Therese's spiritual direction, she wrote her "in order to get something from you, from you who are so close to God"[88]; what she wanted was "the secrets of Jesus" that guided Therese's path to perfection and happiness. Therese responded to Marie's request "to write you my dream and 'my little doctrine.'"[89] The result was a letter of spiritual direction of unparalleled importance in the Theresian corpus. In this letter Therese not only revealed her "totally new" "little way" of spiritual direction, but she also articulated the essence of her spirituality; she shared the result of her search for meaning.

Therese began her direction by reminding Marie of the spiritual equality of all peoples. She admitted "Jesus confides to your little sister" the secrets of happiness, but Marie must "realize He confides these secrets to you too."[90] Then Therese proceeded right to the heart of her spirituality and her spiritual direction: "I understand so well that it is only love which makes us acceptable to God that this love is the only good I [*sic*] ambition." It was Jesus, the "Directors of directors," who guided her to this understanding and showed her the way to achieve this ultimate good: "Jesus deigned to show me the road that leads to this Divine Furnace, and this road is the surrender of the little child who sleeps without fear in its Father's arms."[91] Within these two sentences is the key to understanding Therese's influence and popularity in the twentieth century. She is stating clearly and unambiguously that "love is the only good" capable of making humans happy. Love provides meaning to life. Not only is love the key to human perfection and happiness, but it is also a key that is available to each and every human, even the most insignificant of us, even the sleeping child. Therese's spirituality contained an answer that had been so elusive to those searching for meaning in an age dominated by the politics and philosophies of oppression and despair. It is certainly not a new answer, for we have seen that almost all spiritual directors emphasized love as the answer. It was the way that Therese understood love, and her ability to make her contemporaries see its power and potential, that was significant. Love liberated all people and gave them hope for a better future. Love, simple human love, accomplished both, because it is limitless and, most important, available to all. Through love humans reach their full human potential and attain happiness.

Therese further explained this to Marie by discussing her own personal experience of love's liberating and attainable nature. "To be Your spouse, to be a Carmelite, and by my union with You to be the Mother of souls, should not this suffice me?" Therese rhetorically asked Jesus. But it did not; she wanted more. She wanted to be "the Warrior, the Priest, the Apostle, the Doctor, the Martyr," even a crusader, a papal guard, and a prophet.[92] In fact, Therese desired at one point to engage in "all the actions of all the saints," because she thought that they would bring her to her goal of happiness. She obviously knew that most of these roles she was either not capable of or barred from experiencing. As she sought a resolution to these frustrations in life, Therese read in 1 Corinthians that "all cannot be apostles, prophets, doctors, etc."[93] Still, this answer was not enough, because "it did not fulfill my desires and gave me no peace." So she persevered in her search for happiness. "Without being discouraged, I continued my reading, and this sentence consoled me: 'Yet strive after the better gifts, and I point out to you a yet more excellent way' (1 Cor 12:31; 13:1). And the Apostle explains how all the most perfect gifts are nothing without love. That Charity is the excellent way that leads most surely to God. I finally had rest."[94] Love was her way to happiness.

> Charity gave me the key to my vocation. I understood that if the Church had a body composed of different members, the most necessary and most noble of all could not be lacking to it, and so I understood that the Church had a Heart and that this Heart was Burning With Love. I understood it was Love alone that made the Church's members act, that if Love ever became extinct, apostles would not preach the Gospel and martyrs would not shed their blood. I understood that Love Comprised All Vocations, That Love was Everything, that It Embraced All Times and Places. . . . In a Word, that It was Eternal!
>
> Then, in the excess of my delirious joy, I cried out: O Jesus, my Love—my vocation, at last I have found it—MY VOCATION IS LOVE.
>
> Yes, I have found my place in the Church and it is You, O my God, who has given me this place; in the heart of the Church, my Mother, I shall be LOVE. Thus I shall be everything, and thus my dream will be realized.[95]

For all her reliance on childlike images and themes, Therese was anything but childlike in the analysis of her discovery. Rigorously she penetrated the implications of love being at life's core. She believed that love was responsible for ushering in a new era. "In times past, victims, pure and spotless, were the only ones accepted by the Strong and Powerful God. To satisfy Divine Justice, perfect victims were necessary, but the law of love has succeeded to the law of fear."[96] Just as Jesus replaced the Hebrew Old Law with the New and thus brought salvation to all, so, too, must all people allow Jesus to replace fear with love in their hearts if they desire happiness. This was Therese's directive to her sister Leonine, to enter into the new era of love and to leave behind fear. "We who live under the law of love, how can we fail to put to profit the loving advances our Spouse makes to us? How can we fear One 'who lets himself be held by a hair of our neck?' (Cant 4:9)," Therese asked. "In telling us that a hair can work so great a marvel, He is showing that the smallest actions done for love are the actions which win His heart"—and thus our happiness.[97]

Is it any wonder that so many people during the twentieth century, a century ruled too often by the law of fear in the persons of Lenin, Stalin, Mao, Mussolini, and Hitler, found their will to meaning satiated by Therese's statement? Is it surprising to realize that innumerable of people living under totalitarian regimes and oppressive political systems found Therese's answer liberating and hopeful? One's situation, status, class, or role in society may be limiting or worse but ultimately it is of little consequence, because position cannot fulfill the desire for meaning. Only love can, and every human can choose to love. The option to love is available to every person. It is always a possibility, no matter what situation a person faces. It can never be taken away. In a century when so many people searched for freedom, everyone had the freedom to love. Therese of Lisieux and logotherapists are in agreement here. Logotherapists tell us that the ultimate human freedom is not the freedom from conditions but the freedom to adapt whatever attitude one wants toward one's condition.[98] That Therese fully understood this freedom is clear in her letter to Marie. Here we see that Therese comprehended both her lack of freedom from exterior conditions and her total freedom to choose her interior condition. Therese's decision, freely reached, was to love. In the end, that choice brought her happiness. What "this child asks for is Love," Therese writes about herself. "She knows only one thing: to love You, O Jesus. Astounding works are forbidden to her; she cannot preach the Gospel, shed her blood; but what does it matter since her brothers work in her stead and she, a little child, stays very close to the throne of the King and Queen. She loves in her brothers' place while they do the fighting."[99]

This was the essence of Therese's spiritual direction, her "little way." It is the masses of people with apparently invisible, insignificant jobs who "will strew flowers, [they] will perfume the royal throne with their sweet scent" while others fulfill the more visible role of physical fighting for the king.

"This is how my life will be consumed," Therese herself pledged. Because she was not brilliant, or talented, or a skilled warrior, she had "no other means of proving my love for you other than that of strewing flowers, that is, not allowing one little sacrifice to escape, not one look, one word, profiting by all the smallest things and doing them through love." Yet the menial task of strewing flowers contributed, albeit in a little way, to the glorification of "My Beloved," because it helps create an atmosphere of happiness.[100] Therese thus imparted meaning to even the most trivial human tasks.

Therese admitted this answer was difficult to accept. Once again she offered her personal experience as a way to demonstrate the answer's validity. She knew she was, like so many other people, "too little to perform great actions."[101] This being so, how could an unexceptional, imperfect, little person, like herself, attain the happiness of heaven? "How can a soul as imperfect as mine aspire to the possession of the plentitude of Love? O Jesus, my first and only Friend," Therese exclaimed, "You whom I love Uniquely, explain this mystery to me!"[102] This was the challenge she tried to answer with her spiritual direction: How can the imperfect find happiness? Therese sincerely believed that one's imperfections did not keep one out of heaven. Perfection comes after the fact, because even "the most holy souls will be perfect only in heaven"[103]; something else other than perfection, then, was responsible for happiness. That something was love, and even the imperfect can love. Therese used one of her many effective analogies,[104] this time that of the little bird, to illustrate her point. "I consider myself as a weak little bird, with only a light down as covering. I am not an eagle, But I have only an eagle's eyes and heart," Therese begins. As a little bird desires to fly like the majestic eagle, so, also, Therese wishes to do great things. She is too little for great things, though, and as the bird must realize that "the only thing it can do is raise its little wings, to fly is not within its little power," so must Therese realize her limitations. In her analogy the bird comes not only to accept that but also to acknowledge that it can be perfectly happy just "gazing upon its Divine Sun." When the little bird commits a series of misdeeds in which it wets its wings, "instead of going and hiding away on a corner, to weep over its misery and to die in sorrow, the little bird turns towards its beloved Sun, presenting its wet wings to its beneficent rays" to dry. The bird believes "in the boldness of its full trust that it will acquire in ever greater fullness the love of Him who came to call not the just but sinners" if it simply continues to gaze at the Sun. Even as the little bird seeks the Sun, "when clouds prevent it from seeing a single ray of that Sun, in spite of itself, its little eyes close, its little head is hidden beneath its wing, and the poor thing falls asleep." Again this imperfection does not hinder the bird's happiness, for "when it awaken, it doesn't feel desolate; its little heart is at peace and it begins again its work of love."[105]

I am not proposing that Therese was the first to proclaim that happiness lies in love; the New Testament God is a God of love, and in this God of love resides happiness. In the history of spiritual direction we have seen that most women advocated love as essential to the search for happiness; Hadewijch's poetry is perhaps the most eloquent expression of this approach in spiritual direction. Despite how novel Therese herself thought her discovery of love's role in the fulfillment of human potential to be, it was not. Her originality lies chiefly in her expression. She communicated in simple, contemporary ways the most profound realities of love.

Therese understood spiritual direction to be her personal mission. She was to provide others with specific directions on how to proceed along her little way of love. Therese earnestly believed that "for simple souls there must be no complicated ways,"[106] and so her goal was to help people understand how simple life's mysteries actually were if approached through her "little way." It was a way of love regulated by common sense. In one of Therese's most quoted "little ways" to love, the way of desertion in the face of temptation, we see how she utilized common sense to foster peace and happiness. Therese did not believe in testing the strength of her resolve when tempted by imperfection unless it was absolutely necessary. Instead, she fled if she believed she could be defeated. Therese explained it to Mother Marie de Gonzague by using the metaphor of combat: "My last means of not being defeated in combat is desertion; I was already using this means during my novitiate, and it always succeeded perfectly with me. I wish Mother, to give you an example which I believe will make you smile." Therese then recounted an incident where she was accused of waking up Mother Marie while ill. "I, who felt just the contrary, had a great desire to defend myself. Happily, there came a bright idea into my mind, and I told myself that if I began to justify myself I would not be able to retain my peace of soul. I felt, too, that I did not have enough virtue to permit myself to be accused without saying a word. My last plank of salvation was in flight. No sooner thought than done."[107] Therese thus transformed a supposed vice into a virtue by using the common sense available to everyone.

Desertion was but one of her many little ways to find happiness. Therese took the little annoyances in life—grouchy women, women who made distracting noises in choir, women who unknowingly splashed dirty water in her face—and turned each incident into an opportunity to increase in love and, therefore, happiness. "You can see that I am a very little soul," not a political leader, not an intellectual, not a talented artist, but an average, ordinary person, so "I can offer God only very little things." Even with this minimal approach Therese in her imperfection failed, but she never allowed such imperfection to dismay her. Herein is the secret to the effectiveness and popularity of her spiritual direction: "It often happens that I allow these little sacrifices which give such peace to the soul slip by; this does not discourage me, for I put up with having a little less peace and I try to be more vigilant on another occasion."[108] Therese's spiritual direction exercised such extensive influence over so many people because her little way to perfection and happiness was the simplest. The simplest is always the easiest and the most perfect, a lesson Therese had learned early in her religious life. When she first entered Carmel and had difficulty expressing "what was going on within," her novice mistress Sister Febronie, told her that that was "'because your soul is extremely simple, but when you will be perfect, you will be even more simple; the closer one approaches to God, the simpler one becomes.'"[109] Therese immediately realized how true that was, and she spent the rest of her short life searching for simplicity. Her search resulted in the discovery of simple love and of a simple way to attain it. With that discovery Therese's human potential was realized to the fullest and happiness was hers everlasting: "I shall be LOVE. Thus I shall be everything."[110]

Conclusion

When I began this project I envisioned writing a history of women spiritual directors and men's spiritual direction of women, because I assumed a history solely of women directors would be too sparse to fill a volume. I also assumed there would be lengthy periods when no spiritual direction was given and no women directors existed. My research left me pleasantly surprised. Not only did I find a plethora of women spiritual directors hidden in the sources, but during certain periods my hardest chore was deciding which ones to discuss out of the many possible choices. Certainly this was the situation in the late medieval period, when women mystics flourished, and in the nineteenth century, when novice mistresses abounded. Some readers may be disappointed that I did not include a discussion of more recent and even contemporary women spiritual directors in this survey, but for that I offer no apologies. I have always maintained that a good deal of chronological distance is necessary before the historian can begin work, and I think this rule is especially applicable in the matter of spiritual direction. In clear conscience I may well have included such women as Elizabeth of the Trinity and maybe even Edith Stein, two prominent Carmelite nuns of the first half of the twentieth century, but given the geographic and socially diverse impact of Therese of Lisieux's life and writings, her direction seemed a rather natural endpoint.

I hope two essential points have been established here. The first and by far the most important purpose of this study was to gain an appreciation of the true nature of spiritual direction. It is not merely some kind of authoritative guidance concerning moral behavior or direction given in matters of piety and ritual. It is not a collection of pithy sayings or platitudes. I have argued that the essence of spiritual direction is much deeper. It is the vehicle Christians utilize to address the issues of meaning. As Viktor Frankl has so poignantly argued, life even at its most deprived and perverted level is still capable of providing happiness if that life has meaning; how else could so many people enter the gas chambers of Auschwitz with *Shema Yisrael* or the Lord's Prayer on their lips, upright and content?[1] Because they imparted meaning to their suffering, the sting of death was removed. Instead of misery the experience brought them closer to the ultimate human goal, that of happiness. Meaning was the key. Those who knew *why* could tolerate any *how*. Spiritual direction helps people learn *why*.

Once this function of spiritual direction is realized, its historical significance becomes undeniable. The history of Western civilization in general and Western Christianity in particular has been affected by the work of these directors. Their influential position must be studied as such. This brings us to the second essential fact to emerge from this work, the overwhelming number of women who provided spiritual direction. Women were spiritual directors from the very beginning. At no point in this ministry were women spiritual directors excluded, condemned, or even considered inferior. The only criteria needed to participate in the ministry was a grasp of ultimate realities and the ability to communicate that grasp to others. Throughout the centuries women have met that standard as often as men. It is time we give them the historical respect and attention that is their due.

Notes

Abbreviations

ANF   *The Ante-Nicene Fathers. Translations of the Fathers down to* A.D. *325.* Edited by Alexander Roberts and James Donaldson. Vols. 10. American reprint of the Edinburgh edition. New York: Charles Scribner's Sons, 1903.

AASS   *Acta sanctorum.* Edited by Jean Carnandlt. 2nd ed. Paris: V. Palme, 1863.

JBC   *Jerome Biblical Commentary.* Edited by Raymond E. Brown, Joseph E. Fitzmeyer, and Roland Murphy. Englewood Cliffs, NJ: Prentice-Hall, 1968.

NPNF   *The Nicene and Post Nicene Fathers of the Christian Church.* Second series. Edited by Philip Schaff and Henry Wace. Vols. 14. New York: Charles Scribner's Sons, 1904.

Introduction pp. 1–8

1. Cf. Mary Erler and Maryanne Kowalski, introduction, in *Women and Power in the Middle Ages,* ed. Mary Erler and Maryanne Kowlaski (Athens, GA: University of Georgia Press, 1988), 1–13, for discussion of power so defined.

2. See Carolyn Walker Bynum, *Holy Feast and Holy Fast* (Berkeley: University of California Press, 1987).

3. "Will to power" is associated with Alderian psychology; "will to pleasure" with Freudian psychology; and "will to meaning" with logotherapy.

4. Viktor E. Frankl, *Man's Search for Meaning: An Introduction to Logotherapy,* tr. Isle Lasch, 3rd ed. (New York: Simon and Schuster, 1984), 115.

5. The literature reinforcing this is voluminous. See, for example, Patricia Starck, "Rehabilitative Nursing and Logotherapy: A Study of Spinal Cord Injured Clients," *International Forum for Logotherapy* 4:2 (Fall/Winter 1981), 101–109; George Semno, "Logotherapy in Medical Practice," *International Forum for Logotherapy* 2:2 (Summer/Fall 1979). 12–14; E. Weisskopf-Joelson, "Paranoia and the Will-to-Meaning," *Existential Psychiatry* 1 (1966), 316–320; M. Whidden, "Logotherapy in Prison," *International Forum for Logotherapy* 6:11 (Spring/Summer 1983), 34–39; V. Lieban-Kalma, "Logotherapy: A Way to Help the Learning Disabled Help Themselves," *Academic*

Therapy 19:3 (1984), 262–268; R. M. Holmes, "Alcoholics Anonymous as Group Logotherapy," *Pastoral Psychology* 21 (1970), 30–36; J. C. Crumbaugh and G. L. Carr, "Treatment of Alcoholics with Logotherapy," *International Journal of the Addictions* 14:6 (1979), 847–853; Chris Bennett, "Application of Logotherapy to Social Work Practice," *Catholic Charities Review* 58:3 (March 1974), 1–8.

6. John Paul II, "Encyclical *Laborem exercens.* Addressed by the Supreme Pontiff," *The Evangelist* (September 1981), 18.

7. Ibid., 9. See also Gregory Baum, *Laborem Exercens* (Toronto: University of Toronto, 1983), and George A. Sargent, "Motivation and Meaning: Frankl's Logotherapy in the Work Situation," Ph. D. diss. United States International University, San Diego, 1973. The Opus Dei movement is also most concerned with work as a right. See José Luis Illanes, *On the Theology of Work,* tr. Michael Adams (Dublin: Four Courts Press, 1982).

8. Frankl, *Meaning,* 105.

9. Fredrick Nietzsche, quoted in ibid., 109.

10. Ibid., 132.

11. Viktor E. Frankl, *The Doctor and The Soul,* tr. R. and C. Winston, 2nd ed. (New York: Random House, 1973), 117.

12. Frankl, *Meaning,* 113–114.

13. Ibid., 85.

14. Frankl, *Doctor,* 117.

15. Examples are abundant, especially in John. Life existed only in God, for "through him all things came into being and apart from him nothing came to be. Whatever came to be in him, found Life" (Jn 1:3–4). Jesus "came that they might have life and have it to the full" (Jn 10:10), the fullness of eternal life. Jesus' directions on how to enjoy eternal life were simple: "I solemnly assure you, the man who hears my word and has faith in him who sent me possesses eternal life" (Jn 5:24). See also Jn 11:25–26.

16. *Pilgrimage of the Heart: A Treasury of Eastern Christian Spirituality,* ed. with introd. George A. Maloney (San Francisco: Harper and Row, 1983), 5.

17. Frankl, *Meaning,* 113. The compatibility of the goals of spirituality and logotherapy is apparent when one examines the literature of each discipline. They both believe that one must find meaning in life before one can be happy. Frankl tells us we must all respond to our given situation; Philip Sheldrake, *Spirituality and History: Questions of Interpretation and Method* (New York: Crossroad, 1992), 37, says that spirituality is that which "seeks to express the conscious human response to God that is both personal and ecclesial." Frankl believes that "one should not search for an abstract meaning of life. Everyone has his own specific vocation or mission in life" (*Meaning,* 113); Rowan Williams, *The Wound of Knowledge: Christian Spirituality from the New Testament and St. John of the Cross,* 2nd rev. ed. (Cambridge, MA: Cowley Publications, 1990), 1, argues that spirituality must "be understood in terms of this task: each believer making his or her own that engagement with the questioning at the heart of faith." Moreover, Williams states that

this questioning "is not our interrogation of the data, but its interrogation of us," thus echoing Frankl's statement that "each man is questioned by life; and he can only answer to life by answering for his own life" (*Meaning*, 113–114). Logotherapists help the individual focus "on the future, that is to say, on the meanings to be fulfilled in the future" (Frankl, *Meaning*, 104); historians tell us the spirituality of the first Christians was "predominantly eschatological" (David M. Stanley and Raymond E. Brown, "Aspects of New Testament Thought," *JBC*, 78:72). Frankl, *The Unconscious God* (New York: Simon and Schuster, 1975), 61; and *Meaning*, 122, states that Paul writes that "we speak of these, not in words of human wisdom but in words taught by the Spirit" (1 Cor 2:13), and that Christianity "had none of the persuasive force of 'wise' argumentation, but the convincing power of God" (1 Cor 4–5). R. Panikker defines spirituality as a "way of handling the human condition" (cited in Michael Downey, *Understanding Christian Spirituality* (New York: Paulist Press, 1997), 14); logotherapy helps one deal with "the very essence of human existence" (Frankl, *Meaning*, 114). Downey claims that spirituality pertains to "all dimensions of human life" (*Understanding*, 118); Robert Leslie, *Jesus and Logotherapy* (New York: Abingdon Press, 1965), 60, says logotherapy "is a 'holistic' psychology. Even the chief means advocated to achieve one's end is common to both logotherapy and Christian spirituality, the imitation of models." See Charlotte Bühler, "Basic Theoretical Concepts of Humanistic Psychology" *American Psychologist*, 26 (April 1971), 378, and Phil 3:15–17. There are, of course, also fundamental differences between the logotherapy and spirituality, but these concern their self-imposed limitations. In spirituality, every assertion about the meaning of life must be rooted in the experience of the sacred. In logotherapy the search is not necessarily so restrained.

18. The root of the word "salvation" is found in the Old Testament, from the Hebrew *yesha*, meaning liberty, deliver, prosperity; and in the New Testament, from the Greek *soteria*, meaning deliver, health, save. See James Strong, "salvation," *Strong's Exhaustive Concordance of the Bible* (Nashville, TN: Thomas Nelson Publishers, 1979).

Chapter 1 pp. 9–24

1. Thomas Aquinas, *Summa Theologica*, tr. Fathers of English Dominican Province (Repr.; Westminster, MD: Christian Classics, 1981), PI-II, q. 5, art. 8.

2. Augustine, *Confessions*, tr. R. S. Pine-Coffin (Repr.; Harmondsworth: Penguin Books, 1978), 1:1. ("Quia fecisti nos ad te, et inquietum est cor nostrum, donec requiescat in te.")

3. Raymond E. Brown claims that there is no precedence in Hebrew or Greek for a son calling his mother *woman*, and therefore, its symbolic significance must be acknowledged. The choice is certainly relevant to feminist exegetes pondering Jesus' attitudes toward women. See Raymond E. Brown, tr., *The*

Gospel According to John 1–12, vol. 29: *The Anchor Bible,* ed. William F. Albright and David N. Freedman (New York: Doubleday and Company, 1966), 99; see also Bruce Vawter, "The Gospel According to John," in *JBC,* 63:60; *Navarre Bible: Saint John's Gospel,* tr. B. McCarthy (Dublin: Four Courts Press, 1987), 38–39; and Gary M. Burge, "John," *Evangelical Commentary on the Bible,* ed. Walter A. Elwell (Grand Rapids, MI: Baker Book House, 1989), 849. Each time a woman plays a key role in revealing Jesus' meaning about salvation, he uses the term "woman" to address her, even when he already knows her name.

4. Vawter, "Gospel John," in *JBC,* 63:79. See chapter 7 for Terese of Avila's discussion of the Samaritan woman.

5. Brown, *John 1–12,* 172–173.

6. Cf. also Elizabeth Schüssler Fiorenza, *In Memory of Her* (New York: Crossroad, 1984), 327.

7. Brown, *John 1–12,* appendix I:1.

8. Cf. also Mt 27:55; Lk 24:49; and Jn 19:25.

9. It is difficult to determine how many individual women the gospels are referring to. See Michael Fallon, *The Winston Commentary on the Gospels* (Minneapolis, MN: Winston Press, 1980), 426.

10. Most commentators on Mark summarize the manuscript tradition of the passage and the problems the passage contains. See in particular ancient and medieval commentators quoted in *The Douay - Rheims New Testament,* compiled G. L. Haydock (New York: Edward Dunigan and Brother, 1859; photo reprod. Monrovia, CA: Catholic Treasures, 1991). The *New American Bible* dates the passage in the first century. *Gospel Parallels: A Synopsis of the First Three Gospels,* ed. Burton H. Throckmorton, 4th ed. (Nashville, TN: Thomas Nelson Publishers, 1979), lists the manuscripts in which the passage is found and those in which it is not.

11. Burge, "John," *Evangelical Commentary,* 876.

12. Cf. Susan Haskins, *Mary Magdalen: Myth and Metaphor* (New York: Harcourt Brace and Company, 1993).

13. Thomas Aquinas, *Summa Theologica,* PI-II, q. 5, art. 1.

14. "Save," *Strong's Concordance.*

15. See Mk 5:25–34; and Lk 8: 43–48.

16. See Mk 7:24–30.

17. Cf. Schüssler Fiorenza, *Memory,* 138.

18. See Lk 22:26–27.

19. See Mt 27:55 and Lk 23:49.

20. He repeats this in Acts 22:4.

21. The verse was not as central to Hebrew thought. See discussion in Patricia Ranft, *Women and Spiritual Equality in Christian Tradition* (New York: St. Martin's Press, 1998), 2–9; and 37–41. See also Douglas Hall, *Imaging God* (Grand Rapids, MI: Wm. B. Eerdmans Publishing Co. 1986), 19–20.

22. Rm 8:29; 1 Cor 11:7; 1 Cor 15:49; 2 Cor 3:18; 4:4; Col 1:15; 3:10; Heb 1:3; *Didache,* 5, in *ANF* 7: 379; *First Epistle of Clement,* 33:5, in *ANF*

1:13–14; *Epistle of Ignatius*, in *ANF* 2:110; and *Epistle of Barnabas*, in *ANF* 6:140.

23. Gerhart B. Ladner, *The Idea of Reform* (Cambridge, MA: Harvard University Press, 1959), 9, 26, 54, respectively.

24. Ibid., 62.

25. See Mk 10:17–31 and Lk 18:18–25.

26. The noun *martus* and its derivations are found over two hundred times in the New Testament, primarily in John (eighty-three times) and Acts (thirty-nine times). Cf. Allison A. Trites's definitive work on witness, *The Concept of Witness in New Testament Thought* (Cambridge: Cambridge University Press, 1977), and Patricia Ranft, "The Concept of Witness to the Christian Tradition from Its Origin to Its Institutionalization," *Revue bénédictine* 102:1–2 (1992), 9–23.

27. *Clement*, 5:1, in *ANF* 1:6.

28. Ibid., 9:2, in *ANF* 1:7.

29. Ibid., 17:1, in *ANF* 1:9.

30. Ibid., 7:5, in *ANF* 1:7.

31. Ibid., 19:2, in *ANF* 1:10.

32. Cf. Allison A. Trites, "Martus and Martydom in the Apocalypse: A Semantic Study," *Novum Testamentum* 15:1 (January 1973), 72–80, for the history of the diachronistic semantic change of *martus* from *witness* to *martyr*.

33. *Clement*, 6:1–2, in *ANF* 1:6.

34. See M. Wilma Geisman, "The Use in the Missal and the Breviary of the Roman Liturgy of Judith and Esther of the Old Testament in relation to Christian Womanhood," master's thesis, Notre Dame University, 1956, 18–27, for citations in patristic literature.

Chapter 2 pp. 25–48

1. For a detailed analysis of this thesis, see Ranft, *Women and Equality.*

2. *St. Basil: Letters*, tr. Agnes C. Way (Washington, DC: Catholic University of America Press, 1951), 2.76.

3. Gregory of Nyssa, *The Life of St. Macrina*, tr. W. K. Lowther Clark (London: Society for Promotion of Christian Knowledge, 1916), 962 D-964 A.

4. Ibid., 966 C-D.

5. Ibid., 972 C-D.

6. Ibid., 970 A-B.

7. Ibid., 970 B.

8. Ibid.

9. Ibid., 960 B-C.

10. Ibid., 960 B.

11. Ibid., 1000 A-B.

12. Ibid., 982 A-B.

13. Ibid., 986 D-988 A.

14. Ibid., 988 B.

15. Ibid., 992 D.
16. Ibid., 966 C.
17. Ibid., 970 C-D.
18. Ibid., 972 A.
19. Ibid., 972 B.
20. Ibid., 966 A.
21. Ibid., 982 A.
22. Ibid., 966 D.
23. Ibid., 970 C. Susanna Elm, *Virgins of God* (Oxford: Claredon Press, 1996), 85, argues that Macrina and Emmelia probably did free the family slaves, *manumissio inter amicos.*
24. *Macrina,* 970 C.
25. Cf. Ranft, *Women and Equality,* 53–56.
26. Peter Brown, "The Notion of Virginity in the Early Church," in *Christian Spirituality: Origins to the Twelfth Century,* ed. Bernard McGimn, John Meyerdoff, and Jean Leclercq (New York: Crossroad, 1987), 429.
27. *Macrina,* 984 D.
28. Ibid., 986 A.
29. Ibid., 986 B.
30. Ibid., 988 C.
31. Ibid., 992 B.
32. Ibid., 990 C-D.
33. Cyril of Jerusalem, *Catechetical Lectures,* 4:10, in *NPNF* 7:21.
34. This is the earliest record I have found of such a custom.
35. *Macrina,* 976D-978D.
36. See *NPNF* 5:430–468, published at the end of the nineteenth century. The translation simply quotes the passage from the vita in n. 1, without comment.
37. See also *NPNF* 5:4, prolegomena, in which William Moore, the first translation of Gregory's treatises into English during the nineteenth century, acknowledges Macrina as "the master spirit of the family."
38. *Macrina,* 960 C.
39. Ibid., 966 B.
40. Ibid., 966 C.
41. Ibid., 970 C.
42. Justin Martyr, *Dialogue with Trypho, a Jew,* 8, in *ANF* 1:198.
43. A. M. Malingrey, *Philosphia: Etude d'un groupe de mots dans la litterature grecque* (Paris: n.p., 1961), 207–261.
44. Cf. Elm, *Virgins,* 44–45.
45. Moore, Prolegomena, 2, in *NPNF* 5:8.
46. Gregory of Nyssa, *On the Soul and the Resurrection* in *NPNF,* 5:464.
47. Ibid., 5:465.
48. *Macrina,* 962 C.
49. Ibid., 972 C.
50. Cf. Elm, *Virgins,* 42.

51. *Macrina,* 978 B.
52. Gregory of Nyssa, *On the Soul,* in *NPNF* 5:430.
53. Ibid., 5:431.
54. Ibid., 5:433.
55. Ibid., 5:436.
56. Ibid., 5:467.
57. Ibid., 5:468.
58. Gregory of Nyssa, *On the Formation of Man,* 16, tr. in Henri de Lubac, *Catholicism* (New York: New American Library, 1964), appendix 1. See also translation ("Making of Man") in *NPNF* 5:387–427.
59. Ibid., 22, appendix, 2.
60. See *Macrina,* 970 D.
61. Ibid., 996 D.
62. Gregory of Nyssa, Canticles, 7, quoted in Verna Harrison, "Male and Female in Cappodocian Theology," *Journal of Theological Studies* n.s. 41:2 (October 1990), 445.
63. See pp. 25–26 in text. Letter 204 in *St. Basil Letters,* 2:76.
64. Cf. Kallistos Ware, "The Spiritual Father in St. John Climacus and St. Symeon the New Theologian," *Studia Patristica* 18 (1988), and Harrison, "Male and Female," 443 n.2.
65. Basil, *Homily on Psalms,* tr. in Harrison, "Male and Female," 448–449.
66. See letter 52 and letter 207, *St. Basil Letters;* and citations in Harrison, "Male and Female," 448–449.
67. Harrison, "Male and Female," 447–448.
68. Gregory Nazianzen, *Funeral Oration on His Father,* 8, in *NPNF* 7:257.
69. Ibid., 11, in *NPNF* 7:258.
70. Ibid., 8, in *NPNF* 7:257.
71. Ibid., 9, in *NPNF* 7:257.
72. Ibid., 7, in *NPNF* 7:256.
73. Gregory Nazianzen, *Funeral Oration on Gorgonia,* 4–5, in *NPNF* 7:239.
74. Ibid., 6, in *NPNF,* 7:240.
75. Ibid., 8, in *NPNF* 7:240.
76. Elm, *Virgins,* 102, takes this a step further and argues that the woman-as-man imagery used in *Macrina* is evidence that the vita is not historical but that Gregory's Macrina in a composite figure.
77. Harrison, "Male and Female," 447; see also Elizabeth Clark, "Devil's Gateway and Bride of Christ: Women in the Early Christian World," in Elizabeth Clark, *Ascetic Piety and Women's Faith* (Lewiston, NY: Edwin Mellen Press, 1986), 43; and Kari Vogt, "'Männ-lichwerden'—Aspekte einer urchristlichen Anthropologie," *Concilium* 21 (1985), 434–442.
78. Nazianzen, *Gorgonia,* 19, in *NPNF* 7:243.
79. Ibid., 1, in *NPNF,* 7:238.
80. Ibid., 14, in *NPNF* 7:242. See also letter 197, in *NPNF* 7:461–462 and *Macrina,* 960B.
81. Harrison, "Male and Female," 471.

82. John Chrysostom, letter 6, tr. in Rosemary Rader, "The Role of Celibacy in the Origin and Development of Christian Heterosexual Friendship," Ph.D. diss., Stanford University, 1977, 119.

83. *Life of Olympias,* 9, in *Maenads Martyrs Matrons Monastics,* ed. Ross S. Kraemer (Philadelphia: Fortress Press, 1988), 199.

84. *Olympias,* 10, in ibid.

85. *Olympias,* 13, in ibid., 201.

86. *Olympias,* 15, in ibid., 202.

87. For image of God see 14 and 15 in ibid.; for confessor see 11 and 16, in ibid., 200, 203.

88. *Olympias,* 15, in ibid., 202.

89. Ambrose, *Concerning Virgins,* 3:1.1, in *NPNF* 10:381.

90. Ambrose, Letter 22.1, in *NPNF* 10:436.

91. "You were good enough to write me word that your holiness was still anxious, because I had written that I was so . . ." Ambrose, Letter 41, in *NPNF* 10:445.

92. Jerome, *Book of Illustrious Men,* 135, in *NPNF* 3:384. He also listed among his books the following: "To Eustochium, *On maintaining virginity,* one book of *Epistles to Marcella,* a consolatory letter to Paula, *On the death of a daughter. . . .*"

93. Jerome, Letter 127:5, in *NPNF* 6:254–255.

94. Jerome's prefaces to his translations of books of the Bible in particular reveal how crucial these women were to his role as a scripture scholar. See *NPNF* 6:483–502.

95. Jerome, Letter 127:7, in *NPNF* 6:255.

96. Ibid.

97. Jerome, Letter 29, in Jean Steinmann, *Saint Jerome and His Times,* tr. Ronald Matthew (Notre Dame, IN: Fides Publishers, 1959), 144.

98. Jerome, Letter 77:6–7, in *NPNF* 6:160–161.

99. Ibid., 45:2–3, in *NPNF* 6:59.

100. Ibid., 39:1–2, in *NPNF* 6:49.

101. Ibid., 107:3, in *NPNF* 6:190.

102. Ibid., 107:13, in *NPNF* 6: 195.

103. Ibid., 54:2, in *NPNF* 6:103.

104. Ibid., 108:1, in *NPNF* 6:195.

105. Ibid., 108:4, in *NPNF* 6:196–197.

106. Ibid., 108:27, in *NPNF* 6:210.

107. Ibid., 108:21, in *NPNF* 6:207.

108. Ibid., 108:27, in *NPNF* 6:209–210.

109. Ibid., 108:20, in *NPNF* 6:206.

110. Ibid., 108:27, in *NPNF* 6:209–210: "No mind was ever more docile that was hers."

111. Ibid., 108:15, in *NPNF* 6:203.

112. Ibid., 108:21, in *NPNF* 6:207.

113. Ibid.

114. Ibid., 108:22, in *NPNF* 6:207.
115. Ibid., 108:20, in *NPNF* 6:206.
116. Ibid.
117. Jerome, Letter 23:2, in *NPNF* 6:42.
118. Jerome, Letter 24:1, in *NPNF* 6:42.
119. Jerome, Letter 24:5, in *NPNF* 6:143.
120. Jerome, Letter 78:3, in *NPNF* 6:158.
121. Jerome, Letter 127:2, in *NPNF* 6:253.
122. Augustine, *Confessions,*9:8.
123. Ibid., 5:8; 5:8; 2:3; 1:11; 2:3; and 6:13 respectively.
124. Ibid., 5:9.
125. Ibid., 9:8.
126. Ibid., 3:11.
127. Ibid.
128. Ibid., 3:12.
129. Ibid., 2:3.
130. Ibid., 1:11.
131. Ibid., 6:2.
132. Ibid., 6:1.
133. Ibid.
134. Ibid., 5:9.
135. Ibid.
136. Ibid., 9:9.
137. Ibid.
138. Ibid., 9:7.
139. Ibid., 6:1.
140. Ibid., 9:11.
141. Ibid., 9:9.
142. Ibid.
143. Ibid., 9:10.
144. Ibid.
145. Ibid.

Chapter 3 pp. 49–66

1. William Bark, *The Origins of the Medieval World* (Stanford, CA: Stanford University Press, 1968).
2. For male monasticism, see C. H. Lawrence, *Medieval Monasticism* (London: Longmans, 1984); for female monasticism, see Patricia Ranft, *Women and the Religious Life in Premodern Europe* (New York: St. Martin's Press, 1996).
3. Within generations of the first nuns and monks, clergy were known to become associated with it. One of the more successful early monastic rules was written by Basil, bishop of Caesaria. By the sixth century the intermingling was so common that the hierarchy often went to the monastery to look for new bishops; Gregory the Great was a Benedictine monk.

4. Urban T. Holmes, *A History of Christian Spirituality* (New York: Seabury Press, 1980), 14–47.
5. John Cassian, *Conferences*, tr. Colm Luibheid (New York: Paulist Press, 1985), 1:2.
6. Ibid.
7. *The Sayings of the Desert Fathers: The Alphabetical Collection*, tr. Benedicta Ward, rev. ed. (Kalamazoo, MI: Cistercian Publications, 1984), Syncletica, 1.
8. Cassian, *Conferences*, 1:5.
9. Ibid., 1:7.
10. Ibid., 1:14.
11. Ibid., 1:5.
12. Aphraates, in Juana Raasch, "The Monastic Concept of Purity of Heart and Its Sources, IV: Early Monasticism," *Studia Monastica* 11 (1969), 272–282.
13. Ibid., 308.
14. Ibid., 287.
15. Cassian, *Conferences*, 1:7.
16. *Sayings*, Syncletica, 27.
17. Ibid., Theodora, 2.
18. Ibid., Syncletia, 7.
19. Ibid., Syncletica, 17.
20. Ibid., Syncletica, 24.
21. Ibid. Evagrius Ponticus was of particular importance here. His descriptive psychology exercised tremendous influence in the new Western society, as did his identification of three stages of spiritual development. Cf. Holmes, *Spirituality*, 36–37; and Evagrius Ponticus, *Praktikos. Chapters on Prayer*, tr. John Eudes Bamberger (Spencer, MA: Cistercian Publications, 1970), introduction.
22. *Sayings*, Syncletica, 13.
23. Ibid., Sycletica, 2, 5, 16, 17, and 26.
24. Ibid., 17.
25. Ibid., 16.
26. Ibid., Theodora, 4.
27. Ibid., 5.
28. Ibid., Sarah, 5.
29. *Palladius: The Lausiac History*, tr. Robert T. Meyer (New York: Newman Press, 1964), 34: 3–6.
30. *Sayings*, Syncletica, 12.
31. Irénée Hausherr, *Spiritual Direction in the Early Christian East*, tr. A. P. Gythiel (Kalamazoo, MI: Cistercian Publications, 1990), 287. The Alphabetical Collection is one of several collections of desert wisdom.
32. John Chrysostom, *Letter to Olympias*, in *NPNF*, 1st series (Grand Rapids, MI: Wm. B. Eerdmans, repr. 1978), 9:298.
33. *Palladius*, 34.
34. Ibid., 38: 8–9. The problem was his abandonment of monastic life.
35. Ibid., 46:6.

36. Ibid., 54:4.
37. Ibid., 59:1.
38. Ibid., 63:3.
39. Ibid., 64:1.
40. Gerontius, *The Life of Melanie the Younger,* tr. Elizabeth Clark (Lewiston, NY: Edwin Mellen Press, 1984), 8.
41. Ibid., 58.
42. Ibid., 27.
43. Ibid., 54. Gerontius was a priest who considered Melanie his spiritual guide and who was a close associate of hers. Whether the words recorded by Gerontius actually came from Melanie is, of course, debatable. He does have his own agenda throughout the vita, that of portraying Melanie as a champion of his own theological learning, and, therefore, the vita's narratives about the orthodox nature of her theological debates with heretics should be taken with a grain of salt. There is no compelling reason to doubt that the aristocrats of the city did seek her out to hear her views on spiritual matters. Fifth-century society was awash with various interpretations of life's ultimate mysteries—Donatism, Monophytism, Pelagiansim, Origenism—as the church struggled to identify the orthodox definitions. Surely a renowned, articulate, educated, spiritual leader and monastic founder like Melanie would have been asked her opinions about matters. Unfortunately, having nothing from her own hand, we cannot know precisely what her spirituality was. See Elizabeth Clark, "Piety, Propaganda and Politics in the *Life of Melanie the Younger,*" in *Ascetic Piety,* 61–94.
44. Gerontius, *Melanie,* 29.
45. Ibid., 32.
46. Ibid., 42–43.
47. Claude Peifer, *Monastic Spirituality* (New York: Sheed and Ward, 1966), 162. See also Yves Raquin, "The Spiritual Father: Towards Integrating Western and Eastern Spirituality," in *Abba,* ed. John R. Sommerfeldt (Kalamazoo, MI: Cistercian Publications, 1982), 285: "The abbot or *abbas* had a double function: to be the father of the monks and their spiritual guide."
48. Hausherr, *Spiritual Direction,* 289.
49. Martyrios, *Shirin,* 71, in *Holy Women of the Syrian Orient,* tr. Sebastian P. Brock and Susan A. Harvey (Berkeley: University of California Press, 1987). 179.
50. *Shirin,* 77, in ibid., 180.
51. *Shirin,* 79, in ibid., 181.
52. Theodore the Studite, Letter 117, PG 99, 1548, in Hausherr, *Spiritual Direction,* 296.
53. Theodore the Studite, Letter 115, PG 99, in ibid., 292–293.
54. Ibid.
55. *AASS,* July 6, 613–614, in Hausherr, *Spiritual Direction,* 281.
56. There were, of course, numerous well-known and highly influential hermits in the West during the early Middle Ages, but they were an exception, not

the rule. We must wait until the eleventh century before the West experiences an eremitic movement of note.

57. See JoAnn McNamara, "The Ordeal of Community: Hagiography and Discipline in Merovingian Convents," *Vox Benedictine* 3 (October 1986), 293–326.

58. Venantius Fortunatus, *The Life of the Holy Radedund*, 3, in *Sainted Women of the Dark Ages*, ed. JoAnn McNamara and John E. Halborg (Durham, NC: Duke University Press, 1992), 72.

59. Baudonivia, Bk 2, *The Life of Holy Radegund*, 2, in ibid., 87.

60. Baudonvia, 8, in ibid., 91.

61. Baudonvia, 20, in ibid., 101.

62. Baudonvia, 9, in ibid., 91.

63. Ibid., 92.

64. Baudonvia, 8, in ibid., 91.

65. Baudonvia, 14, in ibid., 96.

66. Baudonvia, 16, in ibid., 98.

67. For reference to the pastoral blessing, see ibid., 18; references to preaching and teaching her flock are in almost half of all the sections.

68. Baudonvia, 20, in *Sainted Women*, 102.

69. *Rule for Nuns*, 48, in Maria McCarthy, *The Rule for Nuns of St. Caesarius of Arles: A Translation with a Critical Introduction* (Washington, DC: Catholic University of America, 1960), 187.

70. "*Caesaria the Insignificant to the Holy Ladies Richild and Radegund*," in *Sainted Women*, 115.

71. Ibid.

72. Ibid., 116.

73. Ibid., 118.

74. Ibid., 116.

75. Ibid., 117–118.

76. Ibid., 116.

77. Bede, *A History of the English Church and People*, tr. Leo Sherley-Price, rev. R. E. Latham (Repr.; Harmondsworth: Penguin Books, 1968), 4:23.

78. In Ireland Ita had a similar record for guidance of future bishops, although her influence was exerted more through education than spiritual direction. Her monastery at Cluain–Credhuil started a renowned school for boys, and here many future church leaders were shaped. She is often called the foster mother of the saints of Ireland. See *AASS*, Jan. 15, 1:1062–1068.

79. Bede, *History*, 4:23.

80. Ibid., 4:19.

81. Bertilla, 2 and 5, in *Sainted Women*, 282 and 285, respectively.

82. Sadalberga, 25, in ibid., 192.

83. Gertrude, 6, in ibid., 226.

84. Rictrude, 15, in ibid., 207.

85. Aldegund, 13, in ibid., 244–245.

86. Aldegund, 13, in ibid., 207.

87. Rudolf of Fulda, *The Life of Saint Leoba,* in *Anglo-Saxon Missionaries in Germany,* tr. C.H. Talbot (New York: Sheed and Ward, 1954), 206.
88. Ibid., 207.
89. Ibid., 208–209.
90. Ibid., 211.
91. Ibid., 214.
92. Ibid.,, 215.
93. Ibid.
94. Ibid., 221 and 214.
95. Ibid.,. 222–223.
96. Ibid., 222.
97. *English Correspondence of Saint Boniface,* tr. Edward Kylie (New York: Cooper Square Publishers, 1966), anonymous, letter 18.
98. Ibid., letter 8.
99. Ibid.
100. Pachomius, quoted in Raasch, "Purity of Heart," 301.
101. *Sayings,* Syncletica, 24.
102. Temptations are also the key for understanding male ascetics' attitudes toward women in the apophthegms. Women are not evil or sinful in themselves; they are merely temptations, and, as such, neutral: "Do you not realize you are a women and that it is through women that the enemy wars against saints?" Ibid., Arsenius, 28.
103. Ibid., Theodora, 2.
104. Ibid., Syncletica, 7, 17, and 15.
105. Ibid., 20.
106. For a concise, accurate history of penance, see Joseph Martos, *Doors to the Sacred* (New York: Doubleday and Company, 1981), 309–364.
107. Cf. G. Mitchell, "Origins of Irish Penance," *Irish Theological Quarterly* 22 (July 1955), 1–14. See particularly p. 11.
108. Literature on the debate about the discovery of the individual is extensive. See Colin Morris, *The Discovery of the Individual* (London: Camelot Press, 1972); Carolyn Walker Bynum, "Did the Twelfth Century Discover the Individual?" in idem, *Jesus and Mother: Studies in the Spirituality of the High Middle Ages* (Berkeley: University of California Press, 1982), 82–109; Jerry Root, *"Space to speke": The Confessional Subject in Medieval Literature* (New York: Peter Lang, 1997).
109. Letter 1 in *Letters of Abelard and Heloise,* tr. Betty Radice (Harmondsworth: Penguin Books, 1974), 115.
110. Root, *Confessional Subject,* 3.

Chapter 4 pp. 67–86

1. Poems in couplet, 1:27–29; 5–7, in *Hadewijch: The Complete Works,* tr. M. Columba Hart (New York: Paulist Press, 1980), 311.
2. Vision 8, in ibid., 284.

3. "Couplets," 3:50–55, in ibid., 322–323.
4. Ibid., 6:1–6, in ibid., 329.
5. Vision 14, in ibid., 304.
6. "Couplets," 10:71–74, in ibid., 337.
7. Vision 11, in ibid., 292.
8. Letter 14, in ibid., 77.
9. Vision 14, in ibid., 305.
10. Letter 13, in ibid., 75.
11. Poems in Stanzas, 30:9.49–52, in ibid., 214.
12. Stanzas 30:10.55–60, in ibid.
13. Stanzas 30:15.89–90, in ibid., 215.
14. Stanzas 25:6.51–54, in ibid., 197.
15. Stanzas 25:5.41–50, in ibid.
16. Vision 9 in ibid., 285–286.
17. Ibid.
18. Letter 4, in ibid., 53–54.
19. Letter 4, in ibid., 55.
20. Letter 5, in ibid.
21. Letter 6, in ibid., 57.
22. "Couplets," 8:1–2, in ibid., 332.
23. Cf. *New Catholic Encyclopedia* (New York: McGraw Hill, 1967), s.v. "Hadewijch" by Norbert de Paepe, claims *minne* is the experience of the soul in relation to God; Stephanus Axters, *The Spirituality of the Old Low Countries,* tr. Donald Attwater (London: Blackfriars, 1959), says it refers to God.
24. Letter 13, in *Hadewijch,* 75.
25. Letter 24, in ibid., 103.
26. Letter 6, in ibid., 57.
27. Letter 2, in ibid., 50.
28. Letter 6, in ibid., 58.
29. Letter 6, in ibid., 59.
30. Vision 11, in ibid., 290.
31. Vision 1, in ibid., 268.
32. "Couplets," 2:93–94, in ibid., 321.
33. Vision 11, in ibid., 291.
34. Vision 11, in ibid., 290.
35. Vision 14, in ibid., 302.
36. Vision 11, in ibid., 292.
37. Vision 11, in ibid., 291.
38. Vision 7, in ibid., 280.
39. Ibid.
40. Letter 2, in ibid., 49.
41. Letter 6, in ibid., 58.
42. Letter 6, in ibid., 59.
43. Vision 1, in ibid., 268.
44. "Couplets," 4:27–30, in ibid., 326.

45. "Couplets," 5, passim, in ibid, 328–329.
46. Stanzas 18:5–14, in ibid., 175. Hadewijch's thoughts here are remarkably similar to Viktor Frankl's thesis in *Meaning*—or rather, vice versa.
47. Stanzas 5:1–7, in ibid., 139.
48. Letter 6, in ibid., 61.
49. Letter 6, in ibid.
50. Letter 2, in ibid., 51.
51. Letter 6, in ibid., 63.
52. Letter 6, in ibid.
53. Letter 5, in ibid., 56.
54. Letter 5, in ibid.
55. Letter 2, in ibid., 49.
56. Letter 2, in ibid., 50.
57. Letter 2, in ibid., 50.
58. Letter 1, in ibid., 47.
59. Letter 2, in ibid., 49.
60. Stanzas 15:3.25, in ibid., 166.
61. Letter 29, in ibid., 115.
62. *Meister Eckhart and the Beguine Mystics,* ed. Bernard McGinn (New York: Continuum, 1994), 1.
63. Much has been written and debated of late about Hadewijch's influence on Eckhart. Most scholars posit that Hadewijch and Hadewijch II (whose poems had once been considered to be Hadewijch's but have not been dealt with here) heavily influenced him. See Louis Bouyer, *Women Mystics* (San Francisco: Ignatius Press, 1993); Jozef van Mierlo, *Hadewijch Mengeldichten* (Brussels: n.p., 1912); A. C. Bouman, "Die literarische Stellung der Dichterin Hadewijch," *Neophilologus* 8 (1923), 270–279; and Saskia Murk-Jansen, "Hadewijch and Eckhart: Amor intellegere est," in McGinn, *Beguine Mystics,* 17–30. Paul A. Dietrich, "The Wilderness of God in Hadewijch II and Meister Eckhart and His Circle," ibid., 31–43, tempers it more by saying that "perhaps it is enough to observe" Eckhart wrote in a world shaped by the "mystical thought and expression of women who lived in a generation or two earlier." (43).
64. Letter 14r, *Letters of Hildegard of Bingen,* tr. Joseph L. Baird and Radd K. Ehrman (New York: Oxford University Press, 1994), 1:53. The clerics of Cologne do, however, admit that they "were greatly astonished that God works through such a fragile vessel" (Letter 15, 1:54). Considering that they were writing to request a copy of a sermon Hildegard had recently preached that had made such a deep impression, the remark seems merely rhetorical.
65. Letter 22, in ibid., 1:74.
66. Letter 35, in ibid., 1:102.
67. Letter 24r, in ibid., 1:83.
68. Letter 51, in ibid., 125.
69. Letter 40, in ibid., 110.

70. Letter 48, in ibid., 1:120.
71. Letter 83, in ibid., 1:180.
72. Letter 84, in ibid, 1:182.
73. Letter 83, in ibid., 1:180.
74. Letter 38, in ibid., 1:106.
75. Letter 50, in ibid., 1:123. (In reply Hildegard chastised her for "the instability of your mind," Letter 50r in ibid., 1:124.)
76. Letter 31, in ibid., 1:94.
77. Letter 40, in ibid., 1:110.
78. Letter 51, in ibid., 1:125–126.
79. Letter 34, in ibid., 1:101.
80. Letter 29, in ibid., 1:93.
81. Letter 15r, in ibid., 1:60.
82. Letter 17, in ibid., 1:68.
83. Letter 52r, in ibid., 1:130.
84. Letter 14r, in ibid., 1:53.
85. Letter 5, in ibid., 1:37.
86. Letter 31r, in ibid., 1:95.
87. Vision 8, in *Hadewijch*, 282.
88. Letter 16r, in *Hildegard*, 1:67.
89. Letter 86 and 85r/a, in ibid., 1:197 and 194.
90. Letter 48r, in ibid., 1:121.
91. Letter 1, in ibid., 1:28.
92. *Hildegard of Bingen: Scivias*, tr. M. Columba Hart and Jane Bishop, intro. Barbara Newman (New York: Paulist Press, 1990), 15.
93. Cf. Gillian T. W. Ahlgren, "Visions and Rhetorical Strategy in the Letters of Hildegard of Bingen," in *Dear Sister*, ed. Karen Cherewatuk and Ulrike Wiethaus (Philadelphia: University of Pennsylvania Press, 1993), 46–63.
94. Sabrina Flanagan, *Hildegard of Bingen, 1098–1179: A Visionary Life* (London: Routledge, 1989), 72.
95. *Hildegard of Bingen: The Book of the Rewards of Life (Liber Vitae Meritorum)*, tr. Bruce W. Hozeski (New York: Oxford University Press, 1994), xvi, states that "*Liber Vitae Meritorum* is a study of human weakness, seemingly inherent in the human consciousness."
96. Ibid., 5:10.15.
97. Ibid., 5:11, 16–17.
98. Ibid., 5:62.81.
99. Ibid.
100. Ibid., 5:63.82.
101. Ibid., 2:1.4.
102. Ibid., 2:5.9.
103. Ibid., 2:9.15.
104. Ibid., 3:13.19.
105. Letter 7, in *Abelard and Heloise*, 204.

106. Ibid., 209.
107. Letter 115, in ibid., 280.
108. Ibid., 279–280.
109. Thomas of Celano, *Legenda,* 36, in *Legend and Writings of S. Clare of Assisi: Introduction: Translation: Studies,* ed. Ignatius Brady (St. Bonaventure, NY: Franciscan Institute, 1953), 43–44.
110. *Legenda,* 11, in ibid., 26.
111. *Legenda,* 27, in ibid., 38.
112. Second letter to Agnes, 8, in *Francis and Clare: The Complete Works,* tr. Regis J. Armstrong and Ignatius C. Brady (New York: Paulist Press, 1982), 195.
113. Letter to Ermentrude, 9 and 16, in ibid., 208.
114. Fourth letter to Agnes, 14–29 (quote, 15), in ibid., 204.
115. Third letter to Agnes, 12–14, in ibid., 200.
116. *Gertrude of Helfta: The Herald of Divine Love,* tr. Margaret Winkworth (New York: Paulist Press, 1993), 1:1.
117. Ibid., 1:3. Winkworth disagrees that this beguine is Mechthild of Magdeburg.
118. Ibid., 1:16.
119. Ibid., 1:12.
120. Ibid., 1:15.
121. Ibid., 2:24.
122. Ibid., 3:30.
123. Epilogue, *Instructions,* in *Angela of Foligno: Complete Works,* tr. Paul Lachance (New York: Paulist Press, 1993), 317.
124. *Instructions,* 36, in ibid., 314.
125. *Ubertinus de Casale: Arbor vitae crucifixae Jesu,* ed. Charles T. Davis, prol. 5, quoted in ibid., 110.
126. *Instructions,* 14, in *Angela,* 267.
127. *Instructions,* 29, in ibid., 290.
128. *Instructions,* 15, in ibid., 269.
129. *Instructions,* 34, in ibid., 301.
130. *Instructions,* 3, in ibid., 234.
131. Ibid., 234–235.
132. Epilogue, in ibid., 317.
133. Cf. third letter to Agnes, 8, in *Francis and Clare,* 199–202.
134. See, for example, *Life of Blessed Juliana of Mont-Cornillon,* tr. Barbara Newman (Toronto: Peregrina Publishing Co., 1989); Thomas de Cantimpré, *Life of Margaret of Ypres,* tr. Margot King (Toronto: Peregrina Publishing Co., 1990); Jacques de Vitry, *Life of Marie d'Oignies,* tr. Margot H. King (Saskatoon: Peregrina Publishing Co., 1989); *Lives of Ida of Nevilles, Lutgard and Alice the Leper,* tr. Martinus Cawley (Lafayette, OR: Our Lady of Guadalupe Abbey, 1987); *The Life of Beatrice of Nazareth, 1200–1268,* tr. Roger De Ganck (Kalamazoo, MI: Cistercian Publications, 1991); and Thomas de Cantimpré, *Supplement to the Life of Marie d'Oignies,* tr. Hugh Feiss (Saskatoon: Peregrina Publishing Co., 1987).
135. Bouyer, *Women Mystics,* 71.

136. Letter to D. Leander of Granada (Seville, 1606), in Gertrude the Great, *Insinuationes Divinae Pietatis,* tr. N. Canteleu (Paris: Fredericum Leonard, 1662), xxxii.

137. Peter of Dacia, *Vita Christinae Stumbelensis,* 2 and 97, cited in John Coakley, "Friars as Confidants of Holy Women in Medieval Dominican Hagiography," in *Images of Sainthood in Medieval Europe,* eds. Renate Blumenfeld-Kosinski and T. Szell (Ithaca, NY: Cornell University Press, 1990), 229–230.

138. *Instructions,* epilogue, in *Angela,* 318.

Chapter 5 pp. 87–106

1. *Catherine of Siena: The Dialogue,* tr. Suzanne Noffke (New York: Paulist Press, 1980), 97.
2. Ibid., 98.
3. Ibid., 99 and 102.
4. Ibid., 99.
5. Ibid., 100.
6. Ibid., 105.
7. Ibid., 102.
8. Ibid., 103.
9. Ibid., 105.
10. Ibid., 6.
11. Ibid., 24.
12. Ibid., 104.
13. Ibid., 105.
14. Ibid., 106.
15. Ibid., 6.
16. Ibid., 47.
17. Ibid., 51.
18. Ibid., 14.
19. Ibid., 23.
20. Ibid., 18.
21. Ibid., 15.
22. Ibid., 20.
23. Ibid., 23.
24. Ibid., 21.
25. Throughout *Dialogue* Catherine refers to Adam's sin, not Eve's. See ibid., 21,115, 126, 134, 135, 140, and 166.
26. Ibid., 21.
27. Ibid., 22.
28. Ibid., 4.
29. Letter 1, *Letters of St. Catherine of Siena,* tr. Suzanne Noffke (Binghamton, NY: Medieval and Renaissance Texts and Studies, 1988), 38–40.
30. Letter 49, in ibid., 148–150.
31. Letter 54, in ibid., 167–168.

32. Letter 60, in ibid., 190.
33. Letter 88, in ibid., 265.
34. Letter 78, in ibid., 239.
35. Letter 70, in ibid., 220.
36. Letter 56 in ibid., 175, and letter 22 in ibid., 85.
37. Letter 59, in ibid., 185. For similar direction of others, see letter 88, in ibid., 265; letter 73, in ibid., 227; letter 70, in ibid., 219; and letter 65, in ibid., 207.
38. *Ancrene Wisse*, 4, in *Anchorite Spirituality: "Ancrene Wisse" and Associated Works*, tr. Anne Savage and Nicholas Watson (New York: Paulist Press, 1991), 114.
39. *Ancrene Wisse*, 8, in ibid., 200.
40. *Ancrene Wisse*, 3, in ibid., 101.
41. *Ancrene Wisse*, 8, in ibid., 206–207. Apparently it was often difficult for the anchorite to balance the need for prayerful solitude and for spiritual directing. Aelred of Rievaulx commented in his letter to his anchorite sister, "How seldom nowadays will you find a recluse alone. At her window will be seated some garrulous old gossip pouring idle tales into her ears," Quote from Aelred of Rievaulx, "A Rule for Life for a Recluse," in *Works of Aelred of Rievaulx*, vol. 1: *Treatises, The Pastoral Prayer* (Spencer, MA: Cistercian Publications, 1971), 1:7.
42. *Ancrene Wisse*, 3, in *Anchorite Spirituality*, 102.
43. Cf. Ann K. Warren, *Anchorites and Their Patrons in Medieval England* (Berkeley: University of California Press, 1985), 18–41.
44. *Book of Margery Kempe*, tr. Susan Dickman, in *Medieval Women's Visionary Literature*, ed. Elizabeth A. Petroff (New York: Oxford University Press, 1986), 316–317.
45. *Book of Margery Kempe*, ed. Sanford B. Meech and Hope Emily Allen (London: EETS, 1940), 42; *Book of Margery Kempe*, tr. B. A. Windeatt (Harmondsworth: Penguin Books, 1985), 77.
46. *Kempe*, in *Medieval Visionary Literature*, 301.
47. *Book of Margery Kempe*, tr. William Butler-Bowden (New York, 1944), 34, quoted in Clarissa W. Atkinson, *Mystic and Pilgrim* (Ithaca, NY: Cornell University Press, 1983), 191.
48. See introduction in Denise N. Baker, *Julian of Norwich's "Showings"* (Princeton, NJ: Princeton University Press, 1994), for discussion of the relationship of the two versions. Scholars agree Julian wrote both, with the short one preceding the long form by some twenty years.
49. *Julian of Norwich: Showings*, tr. Edmund Colledge and James Walsh (New York: Paulist Press, 1978), Short text, 6, and Long text, 8, in ibid., 133 and 191.
50. Short text, 6, in ibid., 134.
51. Ibid.
52. Long text, 9, in ibid., 191.
53. Short text, 6, in ibid., 134.
54. Ibid., 135.

55. Short text, 7, in ibid., 136.
56. Short text, 6, in ibid., 135.
57. Long text, 6, in ibid., 185.
58. Short text, 6, in ibid., 135.
59. Short text, 6, in ibid., 134.
60. The bibliography for the topic is now voluminous. See bibliography in Baker, *Julian "Showings,"* for references.
61. Long text, 80, in *Showings,* Colledge, 335.
62. Long text, 43, in ibid., 255.
63. Long text, 65, in ibid., 308.
64. Long text, 9, in ibid., 192.
65. Long text, 81, in ibid., 337.
66. Long text, 10, in ibid., 196.
67. Long text, 61, in ibid., 302.
68. Long text, 60, in ibid., 297.
69. Long text, 61, in ibid., 301.
70. Long text, 52, in ibid., 280.
71. Johannes von Marienwerder, *The Life of Dorothea von Montau: A Fourteenth Century Recluse,* tr. Ute Stargardt (Lewiston, NY: Edwin Mellen Press, 1997), 2:28.
72 Ibid., 3:1.
73. For the changing relationship among confessors, spiritual directors, and women, see Patricia Ranft, "A Key to Counter Reformation Women's Activism: The Confessor-Spiritual Director," *Journal of Feminist Studies in Religion* 10:2 (Fall 1994), 7–26.
74. For a history of the text used here, see Stargardt's introduction, Marienwerder, *Dorothea,* 1–5.
75. He tells us that "for brevity's sake" he designated himself as B (for *Beichtvater,* confessor) and her second confessor as P (for *prior*). Ibid., 2:27.
76. Ibid., 3:4 and 3:1.
77. Ibid., 1:23.
78. Ibid., 3:21.
79. Ibid., 2:32.
80. Ibid., 3:21.
81. Ibid., 2:29.
82. Ibid., 3:21.
83. Ibid., 2:27.
84. Ibid., 2:28.
85. Quoted in Ranft, "Confessor-Spiritual Director," 15.
86. *Dorothea,* 2:21 and 3:3, are just two examples.
87. Ibid., 2:29.
88. Ibid., 3:3.
89. Ibid., 3:6.
90. *Life,* 16, in *Birgitta of Sweden: Life and Selected Revelations,* ed. M. T. Harris, trans. A. Kezel (New York: Paulist Press, 1990), 75. See also *Life,* 30, in ibid., 79.

91. *Life*, 26, in ibid., 78.
92. *Life*, 30, in ibid., 79.
93. *Life*, 34, in ibid., 80.
94. *Life*, 86, in ibid., 97.
95. *Life*, 84, in ibid., 96.
96. *Life*, 86, in ibid., 96.
97. *Life*, 88, in ibid., 97.
98. *Life*, 23, in ibid., 76.
99. *Life*, 37, in ibid., 81–82.
100. Illuminata Bembo, *Lo specchio di illuminazione*, ed. S. d'Aurizio (Bologna, 1983), 24, in Jeryldene M. Wood, *Women, Art, and Spirituality: The Poor Clares of Early Modern Italy* (Cambridge: Cambridge University Press, 1996), 134.
101. Wood, *Women, Art*, 130.
102. See Serena Spanò Martinelli, "La biblioteca del 'Corpus Domini' bolognese: l'inconsueto spaccato di una cultura monastica femminile," *La Bibliofilía* 88:1 (1986), 1–23.
103. Catherine Vigri of Bologna, *The Seven Weapons of the Spirit*, tr. Joseph R. Berrigan, in *Women Writers of the Renaissance and Reformation*, ed. Katherine M. Wilson (Athens, GA: University of Georgia Press, 1987), prologue. She also repeats her direct authorship in the conclusion.
104. Ibid., 1.
105. Ibid., 3.
106. Ibid., 8.
107. Ibid., 7.
108. See discussion in Richard Kieckhefer, *Unquiet Souls* (Chicago: University of Chicago Press, 1984), 49.
109. Letter 3 (February 23, 1396), in Mary C. Murphy, "Blessed Clara Gambacorta," Ph. D. diss. University of Fribourg, 1928, 113–114.
110. Letter 2 (Christmastide, 1395), in ibid., 111. In Letter 4 (Holy Week, 1396), she wrote, "I have heard that you know how to read. Use this knowledge well" (ibid., 117).
111. Letter 2, in ibid.
112. Ibid., 113.
113. Letter 4, in ibid., 117.
114. Letter 7 (Epiphany, 1397), in ibid., 127.
115. Ibid., 127–129.
116. Letter 15, in ibid., 147. Interestingly, Catherine of Siena wrote two letters of spiritual direction to Clara. For text, see appendix, ibid., 165–172.

Chapter 6 pp. 107–128

1. For Luther's view of confession, see N. N. Rathke, "Die lutherische Auffassung von der Privatbeichte und ihre Bedeutung für das kirchliche Leben der Gegenwart," *Monatschrift für Pastoraltheologie* (1917), 29 ff.; and Michael G. Baylor, *Action and Person: Conscience in Late Scholasticism and the Young*

Luther (Leiden: E.J. Brill, 1977). For a concise history of the sacrament, see Martos, *Doors to Sacred*.

2. Francis de Sales to Madame Brûlart, letter 217, in *Francis de Sales, Jane de Chantal: Letters of Direction*, tr. Péronne M. Thibert (New York: Paulist Press, 1988), 103.

3. *Mémoires pour servir à l'histoire de Port-Royal et à la Vie de la Révérende Mère Marie-Angélique*, 3 vols. (Utrecht, 1742), 1:262, in *The Nuns of Port-Royal as seen in their own narratives*, tr. M. E. Lowndes (London: H. Frowde, 1909), 60.

4. Teresa of Jesus, *Life*, 40 in *The Complete Works of Saint Teresa of Jesus*, tr. and ed. E. Allison Peers, 3 vols. (London: Sheed and Ward, 1957), 1:293. She supports her statement thus: "I have heard the saintly Fray Peter of Alcántra say that, and I have also observed it myself. He . . . gave excellent reasons for this, which there is no point in my repeating here, all in favour of women." Ibid.

5. Electa Arenal and Stacey Schlau, *Untold Sisters: Hispanic Nuns in Their Own Works*, tr. Amanda Powell (Albuquerque, NM: University of New Mexico Press, 1989), 24–25.

6. Mary Ward, "First Instructions to Sisters at St. Omer," in Mary C. E. Chambers, *The Life of Mary Ward (1585–1645)*, ed. Henry James Coleridge, 2 vols. (London: Burns & Oates, 1882–1885), I: 408–411.

7. Chambers, *Mary Ward*, 2:410–411.

8. See *Ascent of Mount Carmel* in *The Collected Works of St. John of the Cross*, tr. Kieran Kavanaugh and Otilo Rodriguez (Washington, DC: ICS Publications, 1979), 160–237; and *Living Flame of God*, in ibid., 621–634.

9. *Ascent*, 4, in ibid., 71.

10. *Living Flame*, 31, in ibid., 621.

11. Ibid., 30, in ibid.

12. Jean J. Surin, *Guide spirituel pour la perfection*, ed. M. de Certeau (Paris: Descleé De Brouwer, 1963). 101.

13. Ana de San Bartolomé, *Letters*, 658–659, in *Untold Sisters*, 76.

14. Domingo Báñez, *Censure*, appendix A, in Carole Slade, *St. Teresa of Avila: Author of a Heroic Life* (Berkeley: University of California Press, 1995), 145.

15. Teresa, *Life*, 13, in *Works of Teresa*, 1:79–80.

16. Ibid., 1:80.

17. Ibid.

18. Ibid., 1:80–81.

19. Ibid., 82. See also, Teresa, *Interior Castle*, 6, 8, in *Works of Teresa*, 2:313; *Book of the Foundations*, 5 and 7, in *Works of Teresa*, 3:19–26 and 3:42–43; and Joel Giallanza, "Spiritual Direction According to St. Teresa of Avila," *Contemplative Review* 12 (Summer 1979), 1–9. John of the Cross learned well from Teresa, who was his early spiritual director, so it is not surprising to see him repeat the essence of her message in his own work: "Besides being learned and discreet, a director should have experience. Although the foundation for guiding a soul to spirit is knowledge and discretion, the director

will not succeed in leading the soul onward in it, when God bestows it, nor will he even understand it, if he has no experience of what true and pure spirit is." *Living Flame,* stanza 3, comm. 30, in *Works of St. John.* 621.

20. Teresa, *Life,* 33, in *Works of Teresa,* 1:226.

21. Ibid., 49, in ibid., 293.

22. Teresa, *Life,* 30, in ibid., 1:195.

23. Ibid., 23, in ibid., I:145–146.

24. Ibid., I: 147.

25. Ibid., I:150.

26. Letter 96 in *Francis, Jane,* 234.

27. Her relationship with Francis de Sales is a classic example.

28. Letter 603 in *Francis, Jane,* 202.

29. Letter 1620, in ibid., 199. See also Letters 1248; 1297; 1318; 1336; 1570; and 1620.

30. Teresa, *Life,* 30, in *Works of Teresa,* I:195.

31. Peers identify it as the convent of Olmeda in Valladolid. See *Works of Teresa,* 3:31 n. 2.

32. Teresa, *Book of Foundations,* 6, in *Works of Teresa,* 3:31–32.

33. Ibid., 8, in *Works of Teresa,* 3:43.

34. Letter 202 to Monsieur Vincent, in *Spiritual Writings of Louise de Marillac: Correspondence and Thoughts,* ed. and tr. Louise Sullivan (Brooklyn, NY: New City Press, 1991), 236.

35. Letter 192, in ibid., 216.

36. Many scholars today, myself included, believe that the spirituality of the French school, like that of Salesan spirituality, was the result equally of women and men. William M. Thompson argues that Madeleine "influenced the theological evolution of Bérulle's spirituality," in addition to playing "a decisive role" in the founding of the Oratory. See *Bérulle and the French School: Selected Writings,* ed. with intro. William M. Thompson, tr. L. M. Glendon (New York: Paulist Press, 1989), 22–23.

37. Madeleine de Saint Joseph, letter 236, in *Bérulle,* 208–209.

38. Ibid., 208.

39. Letter 265 to Monsieur L'abbé de Vaux, in *Writings of Louise,* 267.

40. Letter 196 B to Monsieur Vincent, in ibid., 233.

41. Letter 33 B to Monsieur Vincent, in ibid., 243.

42. Early biographies identify the man in the episode as Vincente Barrón, later ones as Garcia de Toledo. See *Works of Teresa,* 1:234, n.1.

43. Teresa, *Life,* 34, in *Works of Teresa,* 1:235.

44. Ibid., 33, in *Works of Teresa,* 1:226.

45. The trial records are found in Biblioteca comunale di Bologna, MS, 1883. The story is analyzed in Luisa Ciammitti, "One Saint Less: The Story of Angela Mellini, a Bolognese Seamstress (1667–17[?])," in *Sex and Gender in Historical Perspective,* ed. E. Muir and G. Ruggiero (Baltimore: John Hopkins University Press, 1990), 141–176. Quote is from fol. 2v in ibid., 143.

46. Fol. 510r in ibid., 159.

47. Fol. 505r in ibid., 160.
48. Ibid., 161.
49. Letter 37 in *Francis, Jane,* 230.
50. Jeanne Françoise Frémyot de Chantal, *Sa Vie et Ses Oeuvres* (Paris: Plon, 1874–79), 5, 2, 57–58, in Wendy Wright, "Jane de Chantal's Guidance of Women," in *Modern Christian Spirituality: Methodology and Historical Essays,* ed. Bradley C. Hanson (Atlanta, GA: Scholars Press, 1990), 119.
51. Letter 241, in *Francis, Jane,* 241.
52. Letter 398, in ibid., 245.
53. Letter 691, in ibid., 250.
54. Letter 469, in ibid., 247.
55. Teresa, *Book of Foundations,* 4, in *Works of Teresa,* 3:15.
56. Letter 203, in *Francis, Jane,* 221.
57. Jane, *Sa Vie* 4, 1, 168, in Wright, "Jane's Guidance," 122.
58. *Third Mystic of Avila: The self-revelation of Maria Vela, a sixteenth-century Spanish nun,* tr. Frances Parkinson Keyes (New York: Farrar, Straus and Cudahy, 1960), 41–43.
59. Chambers, *Mary Ward,* 1:342–343.
60. Teresa, *Book of Foundations,* 8, in *Works of Teresa,* 3:44.
61. Ibid., 6, in *Works of Teresa,* 3:30–31.
62. Unless, of course, one has already made a vow of obedience to him as confessor-director. This is precisely why the freedom to choose a confessor was so important.
63. Ana de San Bartolemé, *Autobiografía,* 47, I, 295–296, in Arenal and Schlau, *Untold Sisters,* 57.
64. Ibid., 64.
65. Ana de San Bartolemé to Ana de la Ascención, letter August 1621, in ibid., 77.
66. Quoted in Chambers, *Mary Ward,* 1:159.
67. Letter 197, in *Francis, Jane,* 239.
68. Teresa, *Life,* 28 in *Works of Teresa,* 1:185–186.
69. M. Monica, *Angela Merici and Her Teaching Idea (1474–1540)* (New York: Longman Green and Company, 1927), 295.
70. "Rule of St. Angela Merici," 19, in ibid,. 266.
71. The cult was approved by the Congregation of Rites on February 6, 1765, under Clement XIII.
72. Alix Le Clerc, *Relation a la glorie de Dieu et de sa sainte Mère et an salut de mon ame* (Nancy, 1666), quoted in Alfred de Besancenet, *Le Bienheureux Pierre Fournier et la Lorraine* (Paris: René Muffat Libraire, 1864), 44–45.
73. See Ranft, *Women and Religious Life,* 115–116.
74. *Maria Vela,* 41.
75. Ibid., 17–19.
76. Teresa, *Life,* prologue. in *Works of Teresa,* 1:9.
77. Ciammitte, "Angela Mellini," 142–143.

78. Mary Ward, *Autobiography*, preface, in *Till God Will: Mary Ward through her writings*, ed. M. Emmanuel Orchard (London: Darton, Longman and Todd, 1985), 3.

79. Arenal and Schlau, *Untold Stories*, 16. Argument is repeated in idem, "Stratagems of the Strong, Stratagems of the Weak: Autobiographical Prose of the Seventeenth-Century Hispanic Convent," *Tulsa Studies in Women's Literature* 9:1 (Spring 1990), 25–42; and Darcy Donahue, "Writing Lives: Nuns and Confessors as Auto/Biographies in Early Modern Spain," *Journal of Hispanic Philology* 13:3 (1989), 230–239.

80. *Memoires*, in *Nuns of Port-Royal*, 3.

81. "Rule of Angela Merici," 19, in Monica, *Angela*, 266.

82. See Judith Combes Taylor, "From Proselytizing to Social Reform: Three Generations of French Female Teaching Congregations 1600–1720," Ph.D. diss., Arizona State University, 1980, 679–693, for discussion and names of these orders.

83. Mary Poyntz and Winifred Wigmore, *A Briefe Relation of the Holy Life and Happy Death of our Dearest Mother, written c. 1650*, 8–9, in *Till God Will*, 29.

84. *Schola Beatae Mariae*, 1–3 in ibid., 34–35.

85. Letter 41, in *Writings of Louise*, 50.

86. Response to letter 41, in ibid., 51.

87. Jodi Bilinkoff, "Confessors, Penitents, and the Construction of Identities in Early Modern Avila," in *Culture and Society in Early Modern Europe (1500–1800)*, ed. Barbara B. Diefendorf and Carla Hesse (Ann Arbor, MI: University of Michigan Press, 1993), 83–100.

88. Luis de la Puente, *Vida del Padre Baltasar Alvarez* (Madrid: Ediciones Atlas, 1958), 54–55, quoted in Jodi Bilinkoff, *The Avila of Saint Teresa: Religious Reform in a Sixteenth-Century City* (Ithaca, NY: Cornell University Press, 1989), 92.

89. Bilinkoff, "Confessions," 86–96.

90. See Jodi Bilinkoff, "A Saint for a City: Marian de Jesús and Madrid, 1565–1624," *Archiv für Reformationsgeschichte* 88 (1997), 329–330.

91. Teresa, *Life*, 38, in *Works of Teresa*, 1:271.

92. Ibid.

93. Ibid., 1:272.

94. Quoted in Alban Goodier, "St. Teresa and the Dominicans," *Month* (September 1936), 254.

95. Asunciòn Lavrin, "Unlike Sor Juana? The Model Nun in the Religious Literature of Colonial Mexico," *University of Dayton Review* 16:2 (Spring 1983), 75–92.

96. Ward, *Autobiography*, in *Till God Will*, 12.

97. Ibid., 24.

98. M. Jules Michelet, *The Romish Confessional . . . of the Jesuits*, 3rd ed. (Philadelphia: T. B. Peterson, 1851), appendix, 209.

99. Ibid.,. 53–56.

100. See James W. Reites, "Ignatius and Ministry with Women," *The Way* supplement 17 (Summer 1992), 7–19.
101. Michelet, *Romish Confessional*, 56.
102. Ibid., 140–141.
103. Fr. Chiniquy, *The Priest, the Women and the Confessional*, 43rd ed. (New York: Fleming H. Revell Co. 1880), 122–123.
104. Ibid., 128.
105. Ibid., 129–132.
106. René Fülöp-Muller, *The Power and the Secret of the Jesuits*, tr. F. S. Flint (New York: n.p.,1930), vii.
107. Titus Oates, *An exact discovery on the Mystery of Iniquity as it is now practiced among the Jesuits* (1679), was another highly popular anti-Jesuitual treatise published in the nineteenth century (privately printed Edinburgh, 1886, and edited by Edmund Goldsmid).
108. *Secreta Monita Societatis Jesu* (London: L. B. Seeley and Son, 1824), iii-iv.
109. Ibid., 31.
110. Ibid., 41.
111. Ibid., 39. Lewis Owen was a popular seventeenth-century propagandist often used by nineteenth-century authors. See his *The Unmasking of all popish Monks, Friers and Jesuits or, A Treatise of their Geneologie, beginnings, proceedings and present state. Together with some brief observations of their Treasons, Murders, Fornications, Impostures, Blasphemies and sundry other abominable impieties. Written as a cauet or forewarning for Great Britian to take heed in time of these Romish Locusts* (London: F.H. for George Gibs, 1628), 104: "As for the Jesuits, that they are proud, ambitious, aspiring intermedlers . . . they have cunningly wrought, that whatsoever they are, they onely are the generall hearers of all Confessions, diving thereby into the secrets and drifts of all men."

Chapter 7 pp. 129–156

1. Teresa of Jesus, *Conceptions of the Love of God*, 7, in *Works of Teresa*, 2:397–398.
2. Teresa, *Life*, 23, in ibid., 1:152.
3. Teresa, *Way of Perfection*, prol., in ibid., 2:1.
4. See Alison Weber, *Teresa of Avila and the Rhetoric of Femininity* (Princeton, NJ: Princeton University Press, 1990.
5. Teresa, *Way*, prol., in *Works of Teresa*, 2:2.
6. Teresa, *Conceptions*, prol., in ibid., 2:357.
7. See E. Allison Peers's comments in his introduction to *Conceptions* in ibid., 2:353.
8. Ibid., 2:355.
9. Ibid., 2:362.
10. Teresa, *Way*, 3, in ibid., 2:13.
11. Teresa, *Conceptions*, in ibid., 2:362.

12. Cf. Ignatius Loyola, *The Spiritual Exercises,* tr. Thomas Corbishley, 2nd ed. (repr.: Wheathampstead: Anthony Clarke, 1963).

13. For a thorough analysis of Teresa's influence on Francis's spirituality and direction, see M. M. Rivet, *The Influence of the Spanish Mystics on the Works of Saint Francis de Sales* (Washington, DC: Catholic University of America Press, 1941), especially chapter 6. See also letters 216; 223; 280, in *Francis, Jane,* pp. 121; 125; and 126; respectively; and Francis de Sales, *Introduction to a Devout Life. M. J.* (Ratisborn: Frederick Pustat and Co., n.d.), 10.

14. Letter 277, in *Francis, Jane,* 109.

15. See Elizabeth Stopp, *Madame de Chantal* (Westminster, MD: Newman Press, 1963), 106–107; and *Francis, Jane,* 69; 84.

16. *Instruction Given to the Novitiate,* 13, in *Saint Jane Frances Fremyot de Chantal: Her Exhortations, Conferences and Instructions,* rev. ed. (Chicago: Loyola University Press, 1928), 449.

17. Conference 1 in ibid., 145.

18. Elizabeth Rapley, *The Dévotes* (Montreal: McGill-Queen's University Press, 1990), 5.

19. Charles E. Williams, *The French Oratorians and Absolutism, 1611–1641* (New York: Peter Lang, 1989), 53–74.

20. *Works of St. John,* 15–26; Teresa, *Book of Foundations,* 10, in *Works of Teresa,* 3:47.

21. The French School in particular is associated with the phrase "science of mysticism." Bérulle compared his Christo-centric spirituallity with the Copernican Revolution in *Grandeurs;* see Thompson's introduction in *Bérulle,* 5.

22. Teresa, *Way,* in *Works of Teresa,* 2:3.

23. Ibid.

24. Ibid., 2:4–5.

25. Ibid., in *Works of Teresa,* 2:5.

26. Ibid., in *Works of Teresa,* 2:10.

27. Ibid., 2:10–11.

28. Ibid., 2:12.

29. Ibid., 2:11.

30. Ibid., 2:12.

31. Ibid.

32. Ibid., 2:13.

33. Ibid., 2:15.

34. Ibid., in *Works of Teresa,* 2:152. See also ibid., 2:3; 2:10; *Interior Castle,* 7:4, in *Works of Teresa,* 2:351.

35. Ibid., 7:4, in ibid., 2:351. Peers, ibid., 2:189, calls *Interior Castle* "one of the most celebrated books on mystical theology in existence.

36. *Interior Castle,* 1:1, in ibid., 2:201–202.

37. Teresa, *Castle,* 1:1, in ibid., 2:202.

38. *Castle,* 2:1, in ibid., 2:218.

39. *Castle,* 5:3, in ibid., 2:262.

40. *Castle,* 1:1, in ibid., 2:203.
41. *Castle,*1:2, in ibid., 2:208.
42. *Castle,* 1:1, in ibid., 2:204.
43. *Castle,*1:2, in ibid., 2:207.
44. *Castle,* prol., in ibid., 2:200.
45. *Castle,* 1:2, in ibid., 2:209.
46. "Before speaking of the interior life—that is, prayer—" (*Way,* in ibid., 2:16).
47. Ibid.
48. *Way,* in ibid., 2:11–12.
49. Ibid., 2:12.
50. "The other thing you may say is that you are unable to lead souls to God, and have no means of doing so: . . . That is an objection which I have often answered in writing" (*Castle,* 7:4, in ibid., 2:349).
51. *Castle,* 5:3, in ibid., 2: 262–263.
52. *Castle,* 7:4, in ibid., 2:348.
53. *Way,* in ibid., 2:129. In *Conceptions,* 7, in ibid., 2:396, Teresa repeats this: "When the soul is in this state, Martha and Mary never fail, as it were, to work together."
54. The metaphor of water is also highly developed in Teresa's work, particularly in *Way.* Cf. *Way,* in ibid., 2:76–85; and *Life,* in ibid., 1:62–88.
55. *Way,* in ibid., 2:131.
56. *Castle,* 5:1, in ibid., 2:247.
57. *Castle,* 7:2, in ibid., 2:334.
58. Ibid., 2:335.
59. *Exclamations of the Soul to God,* 17, in ibid., 2:420.
60. María de San José, *Book of Recreations,* in Arenal and Schlau, *Untold Sisters,* 96.
61. Ibid., 96–97.
62. Ibid., 99.
63. Ibid., 106.
64. Ibid., 100.
65. Ibid.
66. Ibid., 101.
67. See María de San José, *Escritos espirituales,* ed. Simeón de la Sagrada Familia, 2nd ed. (Rome: Postulación General, 1979); and Arenal and Schlau, *Untold Sisters,* 43.
68. These works are collected under the title *Formacion de novicias y ejercicios de piedad,* in Ana de San Bartolome, *Obras completas de la beata,* ed. Julian Urkiza, 2 vols. (Rome: Edizioni Teresianum, 1981, 1985).
69. Ana de San Bartolome, *Autobiographia,* 47:1. 295–296, in Arenal and Schlau, *Untold Sisters,* 57.
70. *Autobiographia,* 1, 310–311, in ibid., 33.
71. The conflict was addressed in papal bull of 1590. See ibid., 24.
72. For a concise summary of the negotiations, see Williams, *Oratorians,* 53–67.
73. Instruction 1 in *Saint Jane,* 413.

74. Instruction 3, in ibid., 421.
75. Instruction 2, in ibid., 418–419.
76. Ibid., 419.
77. Instruction 10, in ibid., 442.
78. Exhortation 1, in ibid., 19.
79. Instruction 2, in ibid., 415.
80. Instruction 16, in ibid., 454–455.
81. Instruction 17, in ibid., 458.
82. Instruction 20, in ibid., 464.
83. Instruction 18, in ibid., 437.
84. Conference 1, in ibid., 145.
85. Conference 29, in ibid., 238–239.
86. Conference 62, in ibid., 342.
87. *Francis, Jane,* 221.
88. Ibid., 224.
89. Exhortation 7, in *Saint Jane,* 87.
90. Exhortation 2, in ibid., 21.
91. Ibid.
92. Exhortation 6, in ibid., 69.
93. Ibid.
94. Ibid., 70.
95. I willingly admit such a generalization can be questioned on numerous grounds, since there is usually enough evidence to conclude every age is one of change. In this instance I use the statement to imply that once the Christian unity was dissolved in the post-Reformation age, the resulting multiplicity created a new situation.
96. Conference 27, in *Saint Jane,* 233.
97. See Conferences 29, 30, 34, and 36, in ibid. Conferences 27, 28, and 29 deal with simplicity as a virtue, while conferences 44 and 45 discuss simplicity of behavior.
98. Conference 29, in ibid., 238.
99. Ibid., 239.
100. Jeanne Françoise Fréymot de Chantal, *Sa Vie et ses oeuvres* (Paris: Plon, 1874–1879), 8, 5: 605–606.
101. *Francis, Jane,* 236.
102. Ibid., 234.
103. Chantal, *Sa Vie,* 8, 5, 567.
104. Conference 44, in *Saint Jane,* 289.
105. Ibid.
106. Exhortation 4, in ibid., 24.
107. Conference 6, in ibid., 162–163.
108. *Francis, Jane,* 253–254.
109. Ibid., 255.
110. Ibid., 248.
111. Ibid., 263.

112. Ibid., 265.

113. Ibid., 261.

114. Conference 59, in *Saint Jane,* 334.

115. Mary Ward tells us that while still young, "my confessor gave me and recommended me *The Spiritual Combat,* which book was, so to speak, the best master and instructor that I had in spiritual exercises for many years, and one perhaps of the greatest helps, which until now I have enjoyed in the way of perfection." Mary Ward, *Autobiography,* in *Till God Will,* 10. As the title of Lorenzo Scupoli's great classic, *Spiritual Combat,* indicates, much of the spirituality in the post-Reformation era was openly combative. Ironically, Jane de Chantal was also influenced by Scupoli's treatise. See Conference 4, in *Saint Jane,* 157, and Conference 45, in ibid., 294.

116. Letter to Mgr. Albergati, 1620, in *Till God Will,* 24.

117. Ibid., 23.

118. Ibid., 33.

119. *Ratio Instituti in Anglia Historia, 1560–1615,* printed in Chambers, *Mary Ward,* 1:382–383, n. 3.

120. *Till God Will,* 38.

121. *Ratio* in Chambers, *Mary Ward,* 1:375–377. Mary wrote three schemes, the first in 1612 (the *Schola Beatae Mariae*); the second, quoted here, in 1616. They are quite similar; Orchard judges the *Ratio* to be more Ignatian (cf. *Till God Will,* 43). The third scheme was written in 1622, and is quoted next.

122. *Till God Will,* 64.

123. Ibid., 60.

124. Ibid., 56–58.

125. Ibid., 59.

126. Ibid., 60.

127. Father Lohner obtained the collection of sayings from members of the Institute a few decades after Mary Ward's death and inserted them in his biography of her. Cf. Chambers, *Mary Ward,* 1:465–468.

128. Ibid., 1:466.

129. Ibid., 1:467.

130. Ibid., 1:466.

131. "Memorial of the English Clergy to the Holy See," in Chambers, *Mary Ward,* 2:186. The clergy explain further: "The beginnings of this Institute had been received with contempt by many persons as something new and previously unheard of by the Christian world. . . . Yet it made such progress in a very few years, that its disciples have come together into England in great numbers. Wherefore I have deemed it necessary to make the Apostolic See better acquainted with a matter of such moment. . . ." (2:183).

132. Ibid., 1:377.

133. See Ranft, *Women and Religious Life,* 124–128, for a summary of the Institute's history.

134. *De virginibus quibusdam Anglis,* anon., ca. 1614, in *Till God Will,* 48.

135. "Memorial," in Chambers, *Mary Ward,* 2:185.

136. It is my opinion that Mary Ward's dramatic problems and the strong opposition she aroused were rooted in her political maladroitness. She apparently was oblivious to how her frontal attacks on women's spiritual rights alienated many who were in positions to help her.

137. Marie of the Incarnation, *The Relation of 1654,* 6, in *Marie of the Incarnation: Selected Writings,* ed. Irene Mahoney (New York: Paulist Press, 1989), 49.

138. Ibid., 55, and intro., 21; Marie was also influenced by Catherine of Siena.

139. *Relation,* 28, in *Marie: Writings,* 92.

140. *Relation,* 38, in ibid., 110.

141. Ibid., 111.

142. *Relation,* 39, in ibid., 112.

143. *Relation,* 41, in ibid., 116–117.

144. *Relation,* 23, in ibid., 83–84.

145. Letter to Mother Cecile de St. Joseph, in ibid., 273.

146. *Relation,* 49, in ibid., 138.

147. Letter to son (1668), in ibid., 272.

148. Ibid., 271.

149. Letter to son (1650), in ibid., 239.

150. *Relation,* 49, in ibid., 137.

151. *Relation,* 50, in ibid., 139.

152. Letter to son (1668), in ibid., 271.

153. Letter to M. Marie-Gillette Roland, in ibid., 237–238.

154. *Relation,* 50, in ibid., 139.

155. Letter to unnamed friend, quoted in M. Denis Mahoney, *Marie of the Incarnation* (Garden City, NY: Doubleday and Company, 1964), 259.

156. *Writings of Louise,* 1.

157. Louise Sullivan concludes likewise: "With the guidance and support of Vincent, it had devolved on Louise de Marillac to create and sustain this spiritual bond among the young women. . . . The spiritual as well as the professional formation of the sisters was, therefore, largely the task of Louise de Marillac." *Vincent de Paul and Louise de Marillac: Rules, Conferences and Writings,* ed. Francis Ryan and John E. Rybolt, intro. Louise Sullivan (New York: Paulist Press, 1995), intro., 50.

158. Again, Sullivan comments: "Louise would always place a high value on Vincent's guidance but she did not absorb 'Vincentian' spirituality to such an extent that her spiritual journey copied his. As Calvet contends, Vincent de Paul brought her a method rather than a doctrine." Ibid., 42.

159. Letter 129, in *Writings of Louise,* 138.

160. Letter 647b, in ibid., 668.

161. Letter 202, in ibid., 236.

162. Letter 377, in ibid., 406.

163. "Instructions to the Sisters who were sent to Montreuil," A85, in ibid., 773.

164. Letter 302, in ibid., 349.

165. Letter 370, in ibid., 427.

166. *Light,* A2, in ibid., 1.
167. Letter 370, in ibid., 427.
168. Letter 373, in ibid., 426.
169. "Advice requested from Monsieur Vincent," A45, in ibid., 815.
170. Letter 277B, in ibid., 76.
171. Light, A2, in ibid., 1.
172. Ibid.
173. Cf. Edward R. Udovic, "'Caritas Christi Urget Nos': The Urgent Challenges of Charity in Seventeenth Century France," *Vincentian Heritage,* 12:2 (1991), 85–104.
174. Letter 337, in *Writings of Louise,* 385.
175. Letter 414, in ibid., 449.
176. "On the Necessity of Accepting Changes," A66, in ibid., 813–814.

Chapter 8 pp. 157–172

1. Martin Bucer, *On the True Cure of Souls and the Right Kind of Shepherd,* cited in Kenneth Leech, *Soul Friend: A Study of Spirituality* (London: Sheldon Press, 1977), 85.
2. Zwingli, cited in ibid.
3. See Jean-Daniel Benoit, *Calvin, directeur d'ames* (Strasbourg, 1947).
4. *Luther: Letters of Spiritual Counsel,* ed. and tr. Theodore G. Tappert (Philadelphia: Westminster Press, 1955), 22.
5. See Martin Luther, *Table Talk of Martin Luther,* tr. Thomas S. Kepler (Cleveland: World Publishing Co., 1952).
6. Letter to the Christians in Riga, Tallinn, and Tartar, in *Luther: Letters,* 197.
7. The exception to this statement would be the Anglican tradition. Seventeenth-century Anglicans did author some classic works. See George Herbert, *A Priest to the Temple* (London, 1632); Gilbert Burnet, *Discourse of Pastoral Care* (London, 1662); and Jeremy Taylor, *Doctor Dubitantium* (London, 1660).
8. Moravians and later movements led by Mother Ann Lee (Shakers) and Jemima Wilkinson (Universal Friends) were other important radical groups that promoted women spiritual directors. Within contemporary evangelical circles spiritual direction has received some attention. See, for example, Dallas Williard, *In Search of Guidance* (Ventura, CA: Regal, 1984); idem, *The Spirit of Disciplines* (San Francisco: Harper San Francisco, 1991); and Francis A. Schaeffer, *True Spirituality* (Wheaton, IL: Tyndale, 1971).
9. See, for example, William Perkins, *The Whole Treatise of Cases of Conscience* (London, 1602); Immanuel Bourne, *The Godly Man's Guide* (London, 1620); and William Ames, *De Conscientia* (London, 1631).
10. Richard Baxter, *A Breviate of the Life of Margaret . . .* (London, 1681), 67–68, cited in Jacqueline Eales, "Samuel Clarke and the 'Lives' of Godly Women in Seventeenth-Century England," in *Women in the Church,* ed. W. J. Shiels and Diana Wood (Oxford: Basil Blackwood, 1990), 375.

11. Peter Lake, "Feminine Piety and Personal Potency: the 'Emancipation' of Mrs. Jane Ratcliffe," *The Seventeenth Century* 2:2 (July 1987), 158.

12. John Ley, "A petition of piety . . . ," 63, in ibid., 153–154.

13. Ley, 26–27, in ibid., 149.

14. Ley, 27, in ibid., 157.

15. Ley, 106, in ibid.

16. Ley, 100–101, in ibid.

17. Thus Governor Winthrop declared at Anne's trial at Newton in 1637. Rosemary Skinner Keller, "New England Women," in *Women and Religion in America*, vol. 2: *The Colonial and Revolutionary Periods*, ed. Rosemary R. Ruether and Rosemary Skinner Keller (San Francisco: Harper and Row, 1983), intro., 140, claims it was Anne's "radical proposition that spiritual equality before God sanctioned social equality of men and women on earth" that led to her banishment. "You have power over my body but the Lord Jesus hath power over my body and soul," Anne proclaimed at her trial, so if the hierarchy did not acknowledge the superior claim, "you will bring a curse upon you and your posterity." Ibid., doc. 7, 172.

18. Ibid., intro. to doc. 13, 190.

19. Fourth-generation Puritans like Cotton Mather came to believe that women actually were more virtuous than men: "The Grace of God is not confined unto Our Sex alone; so that indeed it is from such a Confinement, that the Instances of it seem to be more Numerous in the other Sex than they are in Ours." Ibid., doc. 3, 339.

20. Ibid., 318.

21. Doc. 2, in ibid., 337–338.

22. Doc. 3, in ibid., 339.

23. Ibid., 340.

24. *A True Testimony from the People of God (Who by the World are Called Quakers)*, (1660), in *A Sincere and Constant Love: An Introduction to the Work of Margaret Fell*, ed. T. H. S. Wallace (Richmond, IN: Friends United Press, 1992), 3.

25. Hester Ann Roe to John Wesley, April 7, 1782, in *Armenian Magazine* 13 (1790), 329, cited in Earl Kent Brown, *Women of Mr. Wesley's Methodism* (New York: Edwin Mellen Press, 1983), 21.

26. Daniel Fristoe, cited in Catherine A. Brekus, *Strangers and Pilgrims: Female Preaching in America, 1740–1845* (Chapel Hill, NC: University of North Carolina Press, 1998), 62.

27. One should not forget that in mid-seventeenth-century England, the question of who was a preacher was not limited to questioning the validity of women preachers. The more widespread, heated issue was over men who claimed to be preachers. This situation led to a bitter literary debate and even the passage of a bill in 1645 that silenced all ministers not properly licensed. Thomas Edward's *Gangraena* was written as part of this debate, and it was this publication that first brought to the public's notice the growing presence of women preachers. See Thomas Edward, *Gangraena* (London,

1646); and Thomas Hall, *The Pulpit Guarded With XX Arguments,* in *Seventeenth Century England: A Changing Culture,* vol. 1: *Primary Sources,* ed. Ann Hughes (London: Open University Press, 1980), 140. Among Quakers Ann Docwra vehemently challenged anyone who called her a preacher: "And to (Francis Bugg's) report, that I am a she-preacher, that he cannot prove, that I ever preached in a public meeting," because "that is not the office in the church that God hath called me to." Anne Docwra, *An Apostate-Conscience Exposed* (London, 1699), in Phyllis Mack, *Visionary Women: Ecstatic Prophecy in Seventeenth-Century England* (Berkeley: University of California Press, 1992), 316.

28. See Susan Mosher Stuard, "Women Witnessing: A New Departure," in *Witnesses for Change,* ed. Elizabeth Potts Brown and Susan Mosher Stuard (New Brunswick, NJ: Rutgers University Press, 1989), 3; Catherine M. Wilcox, *Theology and Women's Ministry in Seventeenth-Century English Quakerism* (Lewiston, NY: Edwin Mellen Press, 1995), 145; Dale A. Johnson, *Women in English Religion 1700–1925* (Lewiston, NY: Edwin Mellen Press, 1983), 40; and Brown, *Women of Methodism,* 113. Although Wilcox says the implications of their beliefs were "without parallel," (145), she also argues that "there were various factors, then, which combined to create an atmosphere in which the Quakers' radical ideas about women could take root" (144). I would deemphasize the radical nature of their ideas and add that Quaker ideas flourished in part because of the long tradition of women's spiritual equality and women spiritual directors that paved the way for them.

29. Anne Laurence, "A Priesthood of She-Believers: Women and Congregations in Mid-Seventeenth-Century England," in *Women in Church,* 363.

30. Christine Trevett, *Women and Quakerism in the Seventeenth Century* (York: Ebor Press, 1991), 13.

31. See Ranft, *Women and Spiritual Equality,* 1–35.

32. For revisionist challenges to this view, see Winthrop S. Hudson, "A Suppressed Chapter in Quaker History," *Journal of Religion* 24 (April 1944), 108–118; Christopher Hill, *World Turned Upside Down* (New York: Penguin Books, 1972); and Barry Reay, *The Quakers and the English Revolution* (New York: St. Martin's Press, 1985).

33. Bonnelyn Young Kunze, *Margaret Fell and the Rise of Quakerism* (Stanford, CA: Stanford University Press, 1994), claims that "Fell's personal power within Quakerism was greater than traditional Quaker historiography has conceded" (230) and that "there is strong evidence that she and Fox were co-founders of equal importance in the Quaker administrative structure" (167). Kunze also says that if we "infer from the word 'mother,' a woman who was an overseer, gentle, meek, and mild, then we have a misnomer in the descriptive title 'mother of Quakerism'" (230).

34. Most literature identifies Margaret Askew by her first husband's name, Fell. Eleven years after Thomas Fell's death she married George Fox.

35. Margaret Fell, *Women's Speaking Justified . . . ,* in *A Sincere Love,* 63.

36. Ibid., 64.

37. Ibid., 72.

38. Letter 291, in *The Works of George Fox*, 8 vols. (Philadelphia: Marcus T. C. Gould, 1831), 8:39–41.

39. Doc. 2, in *Women and Religion*, 283–284. George Fox's sermon on women and the seed has survived, and in it he continues these themes. See *Sermon at Wheeler Street*, in *Early Quaker Writings, 1650–1700*, ed. Hugh Barbour and Arthur O. Roberts (Grand Rapids, MI: Wm. B. Eerdmans, 1973), 501–512.

40. Doc. 2, in *Women and Religion*, 285.

41. This letter was written in 1689; the Swarthmore letter was probably written in 1675. Cited in Mack, *Visionary Women*, 326.

42. Fell, *Women's Speaking*, in *A Sincere Love*, 67.

43. George Keith dedicated an entire work to the ministry of the Samaritan woman, which he held up as a model for all to imitate. "Come hither all you men-preachers of a man-made ministry . . . who cry out against women's preaching and speaking. . . . First of all, she was taught by Christ, by Christ himself; she was taught immediately, and being thus taught, she believed on him, and then she went and preached him. This is an excellent pattern, and example unto all true ministers and preachers of Christ." George Keith, *The Women-Preacher of Samaria* (1674), cited in Wilcox, *Women's Ministry*, 222. See also Thomas Camm, *A Testimony to the Fulfilling the Promise of the Lord relating to such Women . . . become Prophetesses* (London: Thomas Simmons, 1689); and Elizabeth Bathhurst, *The Sayings of Women . . .* (London: Andrew Sowle, 1683).

44. Dorcas Dole, *Once more a Warning to thee O England . . .* (1683), cited in Wilcox, *Women's Ministry*, 211: "When (Mary Magdalen) asked for her Lord in the resurrection of life and power he appeared to her, and endued her with wisdom and power . . . for as the woman fell first into transgression, so in the resurrection Christ first appeared unto her, and sent her to declare unto the men that he was risen."

45. Fell, *Women's Speaking*, 64–67. In an expanded edition of her pamphlet, Margaret presented an even fuller list of biblical women who directed whole communities toward the Truth. Ibid., 73–77.

46. Fox, *Wheeler Street*, in *Quaker Writings*, 505.

47. See George Fox, *To the Men and Womens Monthly and Quarterly Meetings*, in Mack, *Visionary Women*, 286–288; and William Loddington, *The Good Order of Truth Justified: Wherein our Womens Meetings . . . Are Proved Agreeable to Scripture and Sound Reason* (1685), in ibid., 287–288. The debate over the meetings system eventually gave birth to a schism among the Friends. See also Richard T. Vann, *The Social Development of English Quakerism, 1655–1755* (Cambridge, MA: Harvard University Press, 1969).

48. Sarah Fell, *To the Dispersed Abroad, among the Women's meetings every where*, in *Women and Religion*, 283–288, includes care of the poor, needy, widows, and children; approval of marriages; and the collection of testimonies against

the tithes. In Mack's opinion the meetings served as a means of censorship, as a way to encourage and limit prophecy, and as the vehicle through which social services were rendered. It is unfortunate that Mack ignores the work of spiritual direction, for elsewhere she does admit that meetings were where individuals could be "encouraged and protected" and differences resolved. Mack, *Visionary Women,* 280.

49. See Edward Burrough, *The Visitation of the Rebellious Nation of Ireland* (1656), in *Quaker Writings,* 91–92; and Robert Barclay, *Catechism and Confession of Faith* (1673), in ibid., 322–326, for Quakerism's understanding and use of the term.

50. Margaret Hope Bacon, *Mothers of Feminism* (San Francisco: Harper and Row, 1986), 43.

51. *Some Account of the Life and Religious Exercises of Mary Neale, Formerly Mary Peisley* (Dublin, 1795), in Johnson, *Women in English Religion,* 43–44.

52. Sarah Fell, *Dispersed,* in *Women and Religion,* 285–286.

53. Ibid., 287.

54. Letter of October 12, 1680, in Mack, *Visionary Women,* 328–329.

55. Elizabeth Webb, *A Short Account of my Voyage into America with Mary Rogers my Companion,* in ibid., 313.

56. Joan Vokins, *Letter to the Vale of the White Horse Monthly Meeting,* cited in ibid., 343.

57. Alice Hayes, *A Legacy, or Widow's Mite: Left by Alice Hayes, to Her Children and Others* (London, 1723), in ibid., 352.

58. Mary Penington, *Experiences in the Life of Mary Penington,* in *The Hidden Tradition: Women's Spiritual Writings Rediscovered. An Anthology,* ed. Lavinia Byrne (New York: Crossroads, 1991), 107–108.

59. *Life and Letters of Elizabeth L. Comstock* (London: Headley, 1895), 2–3, in ibid., 96.

60. Elizabeth Fry, *Observations on the Visiting, Superintending and Government of Female Prisoners* (London, 1827), 2, 3, in ibid., 96.

61. William Penn, *Just Measures in an Epistle of Peace and Love* (London: Northcott, 1692), 8, in Trevett, *Women and Quakerism,* 85.

62. First published in Albany, 1823; revised 1848.

63. Ibid., in *Gifts of Power: Writings of Rebecca Jackson, Black Visionary, Shaker Eldress,* ed. Jean McMahon Humez (n.c.: University of Massachusetts Press, 1981), 332–333.

64. Ibid., 330.

65. Hannah Cogswell, in *Hidden Tradition,* 140.

66. *Writings of Rebecca,* 277.

67. Letter to Lady Yarborough, in *Susanna Wesley: The Complete Writings,* ed. Charles Wallace (New York: Oxford University Press, 1997), 35.

68. Letter to George Hickes, in ibid., 37.

69. Letter to Samuel, in ibid., 69.

70. Letter to John, in ibid., 150.

71. Letter to Suky, in ibid., 379–380.

72. Ibid., 11. Wallace claims her method "became an early form of 'Methodism' in the autumn of 1729" (144).

73. Journal entries 45 and 43, in ibid., 234 and 229, respectively. She writes extensively about Locke's theory of knowledge and its implication on the spiritual life. Cf. Entries 49 (235–236); 132 (285); 178 (313); 182 (314); 210 (329–330).

74. Entry 26, in ibid., 221.

75. Entry 205, in ibid., 327.

76. Letter to Charles, in ibid., 167–168.

77. Letter to Samuel, in ibid., 50.

78. Letter to John, in ibid., 165.

79. Letter to John, in ibid., 154.

80. In entry 79 Susanna records her scheme for devoting one day each week to one child "and two for Sunday." Ibid., 255.

81. Letter to Samuel, in ibid., 49. Susanna's attitude toward reading was also quite unbending. In her educational treatise she includes this by-law: "8. That no girl be taught to work till she can read very well; and then that she be kept to her work with the same application, and for the same time, that she was held to in reading. This rule also is much to be observed; for putting children to learn sewing before they can read perfectly is the very reason why so few women can read fit to be heard, and never to be well understood." Letter to John, in ibid., 373.

82. Letter to Samuel, in ibid., 48.

83. Letter to John, in ibid., 372.

84. Letter to John, in ibid., 153. See ibid., 156, for possible works she could be referring to.

85. Letter to John, in ibid., 107–109.

86. Letter to John, in ibid., 118.

87. Letter to John, in ibid., 136.

88. Letter to John, in ibid., 161.

89. Ibid., 166.

90. Entry 17, in ibid., 217.

91. Entry 18, in ibid.

92. Entry 28, in ibid., 222.

93. Letter to John, in ibid., 106.

94. Letter to Samuel, in ibid., 50.

95. Entry 17, in ibid., 217.

96. Letter of April 19,1742, in *The Works of John Wesley,* vol. 26: *Letters II, 1740–1755,* ed. Frank Baker (Oxford: Clarendon Press, 1982), 26: 75–76.

97. Letter to Lady Huntingdon, in ibid., 26:115.

98. Letter of May 14, 1757, in Brown, *Women of Methodism,* 35.

99. Ibid., 36.

100. Ibid., 237.

101. Ibid., 42–43.

102. Account of Mrs. Crosby, in ibid., 17.

103. Ibid., 50.

104. Ibid., 127.

105. Ibid., 208.

106. *An Account of the Experience of Hester Ann Rogers* . . . (New York: Lane and Scott, 1849), in *Women and Religion,* 352.

107. Ibid., 355.

108. Brown, *Women and Methodism,* 45.

109. Letter from Lady Huntingdon, in *Works of Wesley,* 73.

110. Brown, *Women and Methodism,* 35.

111. Letter to Countess Delitz, cited in ibid., 39.

112. Samuel Hopkins, *Memoirs of the Life of Mrs. Sarah Osborn* (1799), cited in Mary Beth Norton, "My Resting Reaping Times: Sarah Osborn's Defense of Her 'Unfeminine Activities, 1767,'" *Journal of Women in Culture and Society* 2:2 (1976), 517.

113. Letter to Reverend Joseph Fish, in ibid., 522–529.

114. Ibid., 525.

Chapter 9 pp. 173–192

1. See such handbooks as *Guide to the Catholic Sisterhoods in the United States,* comp. Thomas P. McCarthy (Washington, DC: Catholic University of America Press, 1955), for the religious orders and their stated apostolate. McCarthy lists 915 different orders in the United States alone by mid-twentieth century. All but twenty-one of those orders were founded in the modern period, and the majority were founded in the nineteenth century.

2. Letter 1:2, in Margaret Williams, *St. Madeleine Sophie: Her Life and Letters* (New York: Herder and Herder, 1965), 147.

3. Letter 3:382, in ibid., 434.

4. Letter 1:42, in ibid., 149.

5. Letter 1:49 in ibid., 234.

6. Letter 2:155 in ibid., 242.

7. Letter 3:186 in ibid., 435. Madeleine adds: "Ah, my daughter! What long years it has taken you to accept this!"

8. Letters supp. 2:12 in ibid., 502–503.

9. Letter supp. 1:97, in ibid., 333.

10. Letter 1:98, in ibid., 331.

11. Letter 1:266, in ibid., 438.

12. Letter 1:214, in ibid., 237.

13. Letter 1:70, in ibid., 165.

14. Process of Beautification, in ibid., 498.

15. She was also referring to the Jesuits here. Conferences 1:270, in ibid., 501.

16. Letter 2:168, in ibid., 498.

17. Poitiers journal, in ibid., 497.

18. Letter 1:276, in ibid., 231.

19. Letter 1:365, in ibid., 238.

20. Letter 2:153, in ibid., 241.
21. Conferences 1:16, in ibid., 246.
22. In Margaret Ward, *Life of Saint Madeleine Sophie Foundress of the Society of the Sacred Heart of Jesus,* 2nd ed. (Roehampton, GB: Convent of the Sacred Heart, 1925), 165.
23. Conferences 2:26, in Williams, *Letters,* 441.
24. Journal entry July 19, 1806, in Ward, *Life of Foundress,* 98.
25. Williams, *Letters,* 442.
26. Letter of July 17,1811, in Ward, *Life of Foundress,* 279.
27. Letter 1:83, in Williams, *Letters,* 151.
28. Letter 2:43, in ibid., 152.
29. Letter 1:214, in ibid., 237. Also in Ward, *Life of Foundress,* 282.
30. Letter 2:106, in Williams, *Letters,* 238. Also in Ward, *Life of Foundress,* 286.
31. Williams, *Letters,* 508.
32. Catherine of Siena comes to mind as one spiritual director who did not hesitate to be harsh or even ridicule when needed, even in letters to the pope. See Ranft, *Spiritual Equality,* 187.
33. Letter 2:100, in Williams, *Letters,* 241.
34. Letters 2:417, in ibid., 336.
35. Ibid., 337.
36. Letter 2:179, in ibid., 233–234.
37. Letter 1:80, in ibid., 335.
38. Letter 1:15, in ibid., 158.
39. Mayday letter, cited in Henry James Coleridge, *The Life of Mother Frances Mary Teresa Ball* (London: Burns and Oates, 1881), 156.
40. Undated letter, in ibid., 155.
41. Letter of Dec. 8, 1838, in ibid., 152.
42. Letter of Sept. 18, 1843, in ibid., 167.
43. Undated letter, in ibid., 158.
44. Letter of Apr. 16, 1851, in ibid., 263.
45. Letter of Nov. 14, 1839, in ibid., 153.
46. Undated letter, in ibid., 156.
47. Letter of Dec. 8, 1838, in ibid., 151–152.
48. *The Life of Henriette d'Osseville (In Religion Mother Ste. Marie) Foundress of the Institute of the Faithful Virgin,* ed. John George Macleod (London: Burns and Oates, 1878), 214.
49. Letters 3 and 4, in ibid., 215–216.
50. Letter 8, in ibid., 218. Italics and capitals are as in source. This is also true of Theresa of Lisieux's works that follow.
51. Letter 12, in ibid., 219.
52. On June 10, 1914, Pius X signed the Decree for the Instruction of the Cause for Therese's canonization. In a much-quoted talk, Pius made this remark. See *Story of a Soul: The Autobiography of St. Therese of Lisieux. A New Translation from the Original Manuscripts,* tr. John Clarke (Washington, DC: ICS Publications, 1975), 287.

53. André Combes wrote *Saint Therese and Her Mission: The Basic Principles of Theresian Spirituality*, tr. Alastair Guinar (New York: P. J. Kenedy and Sons, 1995), to refute these criticisms of Therese.

54. Dorothy Day, *Therese* (Notre Dame: Fides Publishers Association, 1960). Noted theologian Michael Novak claims "St. Therese is *the* teacher of the Church about the everyday exercise of *caritas*," and that she had extensive influence on his work. See Michael Novak, "Controversial Engagements," *First Things* 92 (April 1999), 25.

55. *Story*, 207.

56. Ibid.

57. Ibid., 216.

58. Ibid., 160. Therese addressed this part of the manuscript to Pauline. The autobiography was not written as such by Therese but is actually the composite of three different texts with three different addressees. In 1898 the prioress at Lisieux, Mother Marie de Gonzague, ordered the publication of all Therese's writings under the title *Histoire d'une Ame*. In 1973 the publication of Therese's canonization process revealed that this text was really an edited composite of three manuscripts. Manuscript A was written under obedience to her sister Pauline when Pauline (Mother Agnes of Jesus) was prioress, and recorded her childhood memories up until Therese's profession. Manuscript B was written to fulfill Therese's sister Celine (Sister Marie of the Sacred Heart) request for "a souvenir of my retreat" (ibid., 187). Manuscript C was addressed to Mother Marie de Gonzague and records Therese's last years in Carmel.

59. Ibid., 43.

60. Ibid., 13. In mss. C Therese's second sentence stated that Pauline "was the mother entrusted by God with guiding me in the days of my childhood" (ibid., 205).

61. Ibid., 73.

62. Ibid., 57.

63. Ibid., 44–45.

64. Ibid., 74.

65. Letter of May, 1889, in *Saint Therese of Lisieux General Correspondence*, vol. 1: *1877–1890*, tr. John Clarke (Washington, DC: ICS, 1982), 1:563.

66. *Story*, 141.

67. Letter of May 28, 1897, in *Collected Letters of Saint Theresa of Lisieux*, ed. André Combes, tr. F. J. Sheed (New York: Sheed and Ward, 1949), 337.

68. Letter of Sept. 1, 1890, in ibid., 1:658.

69. Letter of May 6, 1884, in ibid., 1:201.

70. Ibid., 201, and *Story*, 76.

71. *Story*, 149.

72. Almost immediately thereafter he was sent to Canada. They corresponded throughout the years, but his letters indicate Therese was more the spiritual director of him than vice versa. See, for example, letter of Oct. 14, 1888, in *Correspondence*, 1:463.

73. *Story,* 150.
74. Ibid., 151.
75. Ibid., 158.
76. Ibid., 159.
77. Ibid., 190.
78. Ibid., 200.
79. Ibid., 234.
80. Ibid., 237–238.
81. Ibid., 238.
82. Ibid., 233.
83. Therese was repeating the words of Jesus as recorded by Margaret Mary in a letter Abbe Belliere, Apr. 25, 1897, in *Collected Letters,* 325.
84. *Story,* 239–240.
85. Ibid., 240.
86. Ibid., 239.
87. This "little way" has in the course of the twentieth century become the most common term to describe Therese's spiritual direction methodology. It is based on her famous prediction made two months before her death, part of which has been quoted above. The entire passage reads thus: "I know that I am going to rest; but I know above all that my mission is about to begin; my mission of making souls love the good God as I love him, to teach my little way to souls. If God answers my requests, my heaven will be spent on earth up until the end of the world. Yes, I want to spend my heaven in doing good on earth." *Story,* 263.
88. Ibid., xiii.
89. Ibid., 189.
90. Ibid., 187.
91. Ibid., 188. "Director of directors," ibid., 151.
92. Ibid., 192.
93. Ibid., 193.
94. Ibid., 194.
95. Ibid.
96. Ibid., 195.
97. *Collected Letters,* 276.
98. Frankl, *Doctor,* 117.
99. Ibid., 196.
100. Ibid.
101. Ibid., 200.
102. Ibid., 197.
103. Ibid., 246.
104. Therese's simple metaphors to explain the profundities of life obviously have contributed both to her popularity among the masses and to her dismissal among the intellectual elite. Perhaps her most famous metaphor was the Divine Elevator; see ibid., 207. See also Letter 170 and 229 in *Collected Letters,* 270–274 and 359; and Combes, *Therese and Her Mission,* passim.

105. *Story,* 198–200.
106. *Story,* 254. While her discovery was not novel, the implications her presentation of love had on the theology of work and the theology of the laity have been called revolutionary by theologian Hans Urs von Balthasar. See Novak, "Engagements," 25.
107. Ibid., 223–224.
108. Ibid., 250.
109. Ibid., 151.
110. Ibid., 194.

Conclusion pp. 193-194

1. Frankl, *Meaning,* 136.

Bibliography

Primary Sources

Ames, William. *De conscientia*. London, 1631.

Ana de San Bartolomé. *Obras completas de la Beata*. Edited by Julian Urikiza. 2 vols. Rome: Edizioni Teresianum, 1985.

Anchorite Spirituality: "Ancrene Wisse" and Associated Works. Translated by Anne Savage and Nicolas Watson. New York: Paulist Press, 1991.

Angela of Foligno. Complete Works. Translated by Paul Lachance. New York: Paulist Press, 1993.

Aquinas, Thomas. *Summa Theologica*. Translated by Fathers of English Dominican Province. Reprint. Westminster, MD: Christian Classics, 1981.

Arenal, Electa and Stacey Schlau. *Untold Sisters: Hispanic Nuns in their Own Works*. Translated by Amanda Powell. Albuquerque, NM: University of New Mexico Press, 1989.

Augustine. *Confessions*. Translated by R. S. Pine-Coffin. Reprint. New York: Penguin Books, 1988.

Barclay, Robert. *Catechism and Confession of Faith*. 1673. In *Early Quaker Writings 1650–1700*. Edited by Hugh Barbour and Arthur O. Roberts. 315–349. Grand Rapids, MI: William B. Eerdmans Publishing Company, 1973.

Bathhurst, Elizabeth. *The Sayings of Women* . . . Shoreditch, 1683.

Baudoniva. *The Life of Holy Radegund*. In *Sainted Women of the Dark Ages*. Edited by JoAnn McNamara and John E. Halborg. 86–105. Durham, NC: Duke University Press, 1992.

Bede. *A History of the English Church and People*. Translated by Leo Sherley-Price. Revised by R. E. Latham. Revised ed. Harmondsworth: Penguin Books, 1968.

Bérulle and the French School: Selected Writings. Edited with introduction by William M. Thompson. Translated by L. M. Glendon. New York: Paulist Press, 1989.

Birgitta of Sweden: Life and Selected Revelations. Edited by M. T. Harris. Translated by A. Kezel. New York: Paulist Press, 1990.

Book of Margery Kempe. Translated by Susan Dickman. 314–329. In *Medieval Women' Visionary Literature*. Edited by Elizabeth A. Petroff. New York: Oxford University Press, 1986.

Book of Margery Kempe. Edited by Sanford B. Meech and Hope Emily Allen. London: EETS, 1940.

Book of Margery Kempe. Translated by B. A. Windeatt. Harmondsworth: Penguin Books, 1985.

Bourne, Immanuel. *The Godly Man's Guide.* London, 1620.

Bucer, Martin. *On the True Cure of Souls and the Right Kind of Shepherd.* London: Sheldon Press, 1977.

Burnet, Gilbert. *The Discourse of Pastoral Care.* London, 1662.

Burrough, Edward. *The Visitation of the Rebellious Nation of Ireland.* London: Giles Calvert, 1656. In *Early Quaker Writings 1650–1700.* Edited by Hugh Barbaour and Arthur O. Roberts. 91–92. Grand Rapids, MI: William B. Eerdmans Publishing Company, 1973.

Camm, Thomas. *A Testimony to the Fulfilling the Promise of the Lord, relating to such Women . . . become Prophetesses.* London: Thomas Simmas, 1660.

Cantimpré, Thomas de. *The Life of Margaret of Ypres.* Translated by Margot H. King. Toronto: Peregrina Publishing Co., 1990.

———. *Supplement to the Life of Marie d'Oignies.* Translated by Hugh Feiss. Saskatoon: Peregrina Publishing Co., 1987.

Cassian, John. *Conferences.* Translated by Colm Luibheid. New York: Paulist Press, 1985.

Catherine of Siena: The Dialogue. Translated by Suzanne Noffke. New York: Paulist Press, 1980.

Catherine Vigri of Bologna. *The Seven Weapons of the Spirit.* Translated by Joseph R. Berrigan. In *Women Writers of the Renaissance and Reformation.* Edited by Katherine M. Wilson. 84–95. Athens, GA: University of Georgia Press, 1987.

Chantal, Jeanne Françoise Frémyot de. *Sa Vie et Ses Oeuvres.* Paris: Plon, 1874–1879.

Chiniquy, Fr. *The Priest, the Women, and the Confessional.* 43rd ed.: New York: Fleming H. Revell Co., 1880.

Collected Letters of Saint Therese of Lisieux. Edited by André Combes. Translated by F. J. Sheed. New York: Sheed and Ward, 1949.

Collected Works of St. John of the Cross. Translated by Kieran Kavanaugh and Otilo Rodriguez. Washington, DC: ICS Publications, 1979.

Complete Works of Saint Teresa of Jesus. Translated and edited by E. Allison Peers. 3 vols. London: Sheed and Ward, 1957.

Cyril of Jerusalem. *Catechetical Lectures. NPNF* 7: 1–160.

Douay-Rheims New Testament. Compiled by G. L. Haydock. New York: Edward Dunegan and Brother, 1859. Photo reprod. Monnovia, CA: Catholic Treasures, 1991.

English Correspondence of Saint Boniface. Translated by Edward Kylie. New York: Cooper Square Publishers, 1966.

Evagrius Ponticus. *Praktikos. Chapters on Prayer.* Translated by John Eudes Bamberger. Spencer, MA: Cistercian Publications, 1970.

Francis and Clare. The Complete Works. Translated by Regis J. Armstrong and Ignatius C. Brady. New York: Paulist Press, 1982.

Francis de Sales. *Introduction to a Devout Life.* Ratisborn: Frederick Pustat and Co., n.d.

Francis de Sales, Jane de Chantal: Letters of Spiritual Direction. Translated by Péronne M. Thibert. New York: Paulist Press, 1988.

Fülöp-Muller, René. *The Power and the Secret of the Jesuits.* Translated by F. S. Flint. New York, 1930.

Gerontius. *The Life of Melanie the Younger.* Translated by Elizabeth Clark. Lewiston, NY: Edwin Mellon Press, 1984.

Gertrude of Helfta: The Herald of Divine Love. Translated by Margaret Winkworth. New York: Paulist Press, 1993.

Gertrude the Great. *Insinuationes Divinae Pietatus.* Translated by N. Cantelau. Paris: Fredericum Leonard, 1662.

Gifts of Power: Writings of Rebecca Jackson, Black Visionary, Shaker Eldress. Edited by Jean McMahon Humez. n.c.: University of Massachusetts Press, 1981.

Gregory Nazianzen. *Funeral Oration on Gorgonia. NPNF* 7:238–245.

———. *Funeral Oration on His Father. NPNF* 7: 255–268.

Gregory of Nyssa. *The Life of St. Marcrina.* Translated by W. K. Lowther Clark. London: Society for Promotion of Christian Knowledge, 1916.

———. *On the Formation of Man.* In Henri de Lubac. *Catholicism* New York: New American Library, 1964.

———. *On the Soul and the Resurrection. NPNF* 5:430–468.

———. *On the Making of Man. NPNF* 5:387–427.

Hadewijch: The Complete Works. Translated by M. Columba Hart. New York: Paulist Press, 1980.

Herbert, George. *A Priest to the Temple.* London, 1632.

Hidden Tradition: Women's Spiritual Writings Rediscovered. An Anthology. Edited by Lavinia Bynne. New York: Crossroad, 1991.

Hildegard of Bingen: The Book of the Rewards of Life (Liber Vitae Meritorum). Translated by Bruce W. Hazeski. New York: Oxford University Press, 1994.

Hildegard of Bingen: Scivias. Translated by M. Columba Hart and Jane Bishop. Introduction by Barbara Newman. New York: Paulist Press, 1990.

Jerome. *Book of Illustrious Men. NPNF* 3:380–384.

John Paul II. "Encyclical *Laborem exercens.* Addressed by the Supreme Pontiff." *The Evangelist,* (September 1981).

Julian of Norwich: Showings. Translated by Edmund Colledge and James Walsh. New York: Paulist Press, 1978.

Justin Martyr. *Dialogue with Trypho, a Jew. ANF* 1:194–270.

Legend and Writing of S. Clare of Assisi: Introduction: Translation: Studies. Edited by Ignatius Brady. St. Bonaventure, NY: Franciscan Institute, 1953.

Letters of Abelard and Heloise. Translated by Betty Radice. London: Penguin Books, 1974.

Letters of Hildegard of Bingen. Translated by Joseph L. Baird and Radd K. Ehrman. New York: Oxford University Press, 1994.

Letters of St. Catherine of Siena. Translated by Suzanne Noffke. Binghamton, NY: Medieval and Renaissance Texts and Studies, 1988.

Letters of Saint Teresa of Jesus. Edited and translated by E. Allison Peers. 2 vols. London: Burn Oates and Washbourne, 1951.

Life of Beatrice of Nazareth. Translated by Roger De Ganck. Kalamazoo, MI: Cistercian Publications, 1991.

Life of Blessed Juliana of Mont-Cornillon. Translated by Barbara Newman. Toronto: Peregrina Publishing Co., 1989.

Lives of Ida of Nevilles, Lutgard and Alice the Leper. Translated by Martinus Cawley. Lafayette, OR: Our Lady of Guadalupe Abbey, 1987.

Loyola, Ignatius. *The Spiritual Exercises.* Translated by Thomas Corbishley, Reprint. 2nd ed. Wheathampstead: Anthony Clarke, 1979.

Luther: Letters of Spiritual Counsel. Edited and translated by Theodore G. Tappert. Philadelphia: Westminster Press, 1955.

Luther, Martin. *Table Talk of Martin Luther.* Translated by Thomas S. Kepler. Cleveland: World Publishing Co., 1952.

Maenads Martyrs Matrons Monastics. Edited by Ross S. Kraemer. Philadelphia: Fortress Press, 1988.

María de San José. *Escritos espirituales.* Edited by Simeón de la Sagrada Familia. Rome: Postulaci}n General, 1979.

Marie of the Incarnation: Selected Writings. Edited by Irene Mahoney. New York: Paulist Press, 1989.

Marienwerder, Johannes von. *The Life of Dorothea von Montau: A Fourteenth Century Recluse.* Translated by Ute Stargardt. Lewiston, NY: Edwin Mellen Press, 1997.

Martyrios. *Shirin.* In *Holy Women of the Syrian Orient.* Translated by Sebastian P. Brock and Susan A. Harvey. Berkeley: University of California Press, 1987.

McCarthy, Maria. *The Rule for Nuns of St. Caesarius of Arles: A Translation with a Critical Introduction.* Washington, DC: Catholic University of America Press, 1960.

Medieval Women's Visionary Literature. Edited by Elizabeth A. Petroff. New York: Oxford University Press, 1986.

Michelet, M. Jules. *The Romish Confessional or the Auricular Confession and Spiritual Direction of the Romish Church, its History, Consequences and policy of the Jesuits.* 3rd ed.: Philadelphia: T. B. Peterson, 1851.

Navarre Bible: Saint John's Gospel. Translated by B. McCarthy. Dublin: Four Courts Press, 1987.

Nuns of Port-Royal as seen in their own narratives. Translated by M. E. Lowndes. London: H. Frowde, 1909.

Oates, Titus. *An exact discovery on the Mystery of Iniquity as it is now practiced among the Jesuits.* 1679. Privately printed Edinburgh, 1886.

Owen, Lewis. *The Unmasking of All Popish Monks, Friers and Jesuits or, A Treatise of Their Geneologie, Beginnings, Proceedings and Present State. Together with Some Brief Observations of Their Treasons, Murders, Fornications, Impostures, Blasphemies and Sundry Other Adominable Impieties. Written as a Cauet or Forewoarning for Great Britain to Take Heed in Time of These Romish Locusts.* London: F. H. for George Gibs, 1628.

Palladius: The Lausiac History. Translated by Robert T. Meyer. New York: Newman Press, 1964.

Perkins, Wiliam. *The Whole Treatise of Cases of Conscience.* London, 1602.

Pilgrimage of the Heart: A Treasury of Eastern Christian Spirituality. Edited by George A. Maloney. San Francisco: Harper and Row, 1983.

Rudolf of Fulda. *The Life of Saint Leoba.* In *Anglo-Saxon Missionaries in Germany.* Translated by C. H. Talbot. New York: Sheed and Ward, 1954.

Sayings of the Desert Fathers: The Alphabetical Collection. Translated by Benedicta Ward. Rev. ed. Kalamazoo, MI: Cistercian Publications, 1984.

St. Basil: Letters. Translated by Agnes C. Way. Washington, DC: Catholic University of America Press, 1951.

Saint Jane Frances Fremyot de Chantal: Her Exhortations, Conferences, and Instructions. Rev. ed. Chicago: Loyola University Press, 1928.

Saint Therese of Lisieux General Correspondence. Vol. 1: *1877–1890.* Translated by John Clarke. Washington, DC: ICS Publications, 1982.

Secreta Monita Societatis Jesu. London: L. B. Seeley and Son, 1824.

Seventeenth Century England: A Changing Culture. Vol. 1: *Primary Sources.* Edited by Ann Hughes. London: Open University Press, 1980.

Sincere and Constant Love: An Introduction to the Works of Margaret Fell. Edited with and introduction by T. H. S. Wallace. Richmond, IN: Friends United Press, 1992.

Spiritual Writings of Louise de Marillac: Correspondence and Thoughts. Edited and translated by Louise Sullivan. Brooklyn, NY: New City Press, 1996.

Story of a Soul: The Autobiography of St. Therese of Lisieux. A New Translation from the Original Manuscripts. Translated by John Clarke. Washington, DC: ICS Publications, 1975.

Susanna Wesley: The Complete Writings. Edited by Charles Wallace. New York: Oxford University Press, 1997.

Surin, Jean J. *Guide spirituel pour la perfection.* Edited by M. de Certeau. Paris: Desclée De Brouwer, 1963.

Taylor, Jeremy. *Doctor Dubitantium.* London, 1660.

The Third Mystic of Avila: The Self-Revelation of Maria Vela, a Sixteenth-Century Spanish Nun. Translated by Frances Parkinson Keyes. New York: Farrar, Strauss and Cudahy, 1960.

Till God Will: Mary Ward through Her Writings. Edited by M. Emmanuel Orchard. London: Darton, Longman Todd, 1985.

Venantius Fortunatus. *The Life of the Holy Radegund.* In *Sainted Women of the Dark Ages.* Edited by JoAnn McNamara and John E. Halborg. 70–86. Durham, NC: Duke University Press, 1992.

Vincent de Paul and Louise de Marillac: Rules, Conferences, and Writings. Edited by Francis Ryan and John E. Rybolt. Introduction by Louise Sullivan. New York: Paulist Press, 1995.

Vitry, Jacques de. *The Life of Marie d'Oignies.* Translated by Margot H. King. Toronto: Peregrina Publishing Co., 1989.

Williams, Margaret. *St. Madeleine Sophie: Her Life and Letters.* New York: Herder and Herder, 1965.

Williams, Rowan. *The Wound of Knowledge: Christian Spirituality from the New Testament to St. John of the Cross.* 2nd rev. ed. Cambridge, MA: Cowley Publications, 1990.

Women and Religion: A Feminist Sourcebook of Christian Thought. Edited by Elizabeth Clark and Herbert Richardson. New York: Harper and Row, 1977.

Women and Religion in America. Vol. 2: *The Colonial and Revolutionary Periods.* Edited by Rosemary R. Ruether and Rosemary Skinner Keller. San Francisco: Harper and Row, 1983.

Works of Aerled of Rievaulx. Vol. 1: *Treatises, The Pastoral Prayer.* Spencer, MA: Cistercian Publications, 1971.

Works of George Fox. 8 vols. Philadelphia: Marcus T. C. Gould, 1831.

Works of John Wesley. Vol. 26: *Letters II, 1740–1755.* Edited by Frank Baker. Oxford: Clarendon Press, 1982.

Secondary Sources

Ahlgren, Gillian T. W. "Visions and Rhetorical Strategy in the Letters of Hildegard of Bingen." In *Dear Sister.* Edited by Karen Cherewatuk and Ulrike Wiethaus. 46–63. Philadelphia: University of Pennsylvania Press, 1993.

Arenal, Electa, and Stacey Schlau. "Stratagems of the Strong, Stratagems of the Weak: Autobiographical Prose of the Seventeenth-Century Hispanic Convent." *Tulsa Studies in Women's Literature* 9:1 (Spring 1990), 25–42.

Atkinson, Clarissa W. *Mystic and Pilgrim.* Ithaca, NY: Cornell University Press, 1983.

Axters, Stephanus. *The Spirituality of the Old Low Countries.* Translated by Donald Attwater. London: Blackfriars, 1959.

Bacon, Margaret Hope. *Mothers of Feminism.* San Francisco: Harper and Row, 1986.

Baker, Denise. *Julian of Norwich's "Showings": From Vision to Book.* Princeton, NJ: Princeton University Press, 1994.

Bark, William Carroll. *Origins of the Medieval World.* Stanford, CA: Stanford University Press, 1958.

Baum, Gregory. *Laborer Exercens.* Toronto: University of Toronto, 1983.

Baylor, Michael G. *Action and Person: Conscience in the Late Scholasticism and the Young Luther.* Leiden: E. J. Brill, 1977.

Bennet, Chris. "Application of Logotherapy to Social Work Practice." *Catholic Charities Review,* 58:3 (March 1974), 1–8.

Benoit, Jean Daniel. *Calvin, directeur d'ames.* Strasbourg, 1947.

Besancenet, Alfred de. *Le Bienheureux Pierre Fournier et la Lorraine.* Paris: René Muffat Libraire, 1864.

Bilinkoff, Jodi. *The Avila of Saint Teresa: Religious Reform in a Sixteenth-Century City.* Ithaca, NY: Cornell University Press, 1989.

———. "Confessors, Penitents, and the Construction of Identities in Early Modern Avila." In *Culture and Society in Early Modern Europe (1500–1800).* Edited by Barbara Diefendorf and Carla Hesse. 83–100. Ann Arbor, MI: University of Michigan Press, 1993.

———. "A Saint for a City: Mariana de Jesús and Madrid, 1565–1624." *Archiv für Reformationsgeschichte* 88 (1997), 322–337.

Bouyer, Louis. *Women Mystics.* San Francisco: Ignatius Press, 1993.

Bouman, A. C. "Die Literarische Stellung der Dichterin Hadewijch." *Neophilologus* 8 (1923), 270–279.

Brekus, Catherine A. *Strangers and Pilgrims: Female Preaching in America, 1740–1845.* Chapel Hill, NC: University of North Carolina Press, 1998.

Brown, Earl Kent. *Women of Mr. Wesley's Methodism.* New York: Edwin Mellen Press, 1983.

Brown, Peter. "The Notion of Virginity in the Early Church." In *Christian Spirituality: Origins to the Twelfth Century.* Edited by Bernard McGinn, John Meyerdoff, and Jean Leclercq. 427–443. New York: Crossroad, 1987.

Brown, Raymond E. *The Gospel According to John 1–12.* Vol. 29: *The Anchor Bible.* Edited by William F. Albright and David N. Freedman. New York: Doubleday and Company, 1966.

Bühler, Charlotte. "Basic Theoretical Concepts of Humanistic Psychology." *American Psychologist* 26 (April 1971), 378–386.

Burge, Gary M. "John." *Evangelical Commentary on the Bible.* Edited by Walter A. Elwell. Grand Rapids, MI: Baker Book House, 1989.

Bynum, Carolyn Walker. *Holy Feast and Holy Fast.* Berkeley: University of California Press, 1987.

————. *Jesus as Mother. Studies in the Spirituality of the High Middle Ages.* Berkeley: University of California Press, 1982.

Ciammitti, Luisa. "One Saint Less: The Story of Angela Mellini, a Bolognese (1667–17[?])." In *Sex and Gender in Historical Perspective.* Edited by E. Muir and G. Ruggiero. 141–176. Baltimore: John Hopkins University Press, 1990.

Chambers, Mary C. E. *The Life of Mary Ward, 1585–1645.* 2 vols. Edited by Henry James Coleridge. London: Burns and Oates, 1882–1885.

Clark, Elizabeth. "Devil's Gateway and Bride of Christ: Women in the Early Christian World." In *Ascetic Piety and Women's Faith.* 23–60. Lewiston, NY: Edwin Mellen Press, 1986.

————. "Piety, Propaganda and Politics in the Life of Melanie the Younger." In *Ascetic Piety and Women's Faith.* 61–94. Lewiston, NY: Edwin Mellen Press, 1986.

Coakley, John. "Friars as Confidants of Holy Women in Medieval Dominican Hagiography." In *Images of Sainthood in Medieval Europe.* Edited by Renate Blumenfeld-Kosinski and T. Szell. 220–246. Ithaca, NY: Cornell University Press, 1990.

Coleridge, Henry James. *The Life of Mother Frances Mary Teresa Ball.* London: Burns and Oates, 1881.

Combes, André. *Saint Therese and Her Mission: The Basic Principles of Theresian Spirituality.* Translated by Alastair Guinar. New York: P. J. Kenedy and Sons, 1955.

Crumbaugh, J. C. and G. L. Carr. "Treatment of Alcoholics with Logotherapy." *International Journal of the Addictions* 14:6 (1979), 847–853.

Day, Dorothy. *Therese.* Notre Dame, IN: Fides Publishers Association, 1960.

Dietrick, Paul A. "The Wilderness of God in Hadewijch II and Meister Eckhart and His Circle." In *Meister Eckhart and the Beguine Mystics.* Edited by Bernard McGinn. New York: Continuum, 1994.

Donahue, Darcy. "Writing Lives: Nuns and Confessors as Auto/Biographies in Early Modern Spain." *Journal of Hispanic Philology* 13:3 (1989), 230–239.

Downey, Michael. *Understanding Christian Spirituality.* New York: Paulist Press, 1997.

Eales, Jacqueline. "Samuel Clarke and the 'Lives' of Godly Women in Seventeenth-Century England." In *Women in the Church.* Edited by W. J. Shiels and Diana Wood. 365–376. Oxford: Basil Blackwood, 1990.

Edwards, Thomas. *Gangraena* (London, 1646).

Elm, Susanna. *Virgins of God.* Oxford: Clarendon Press, 1996.

Fallow, Michael. *The Winston Commentary on the Gospels.* Minneapolis, MN: Winston Press, 1980.

Flanagan, Sabrina. *Hildegard of Bingen, 1098–1179: A Visionary Life.* London: Routledge, 1989.

Fiorenza, Elizabeth Schüssler. *In Memory of Her.* New York: Crossroad, 1984.

Frankl, Viktor E. *The Doctor and the Soul.* Translated by R. and C. Winston. 2nd ed. New York: Random House, 1973.

———. *Man's Search for Meaning: An Introduction to Logotherapy.* Translated by Isle Lasch. 3rd ed. New York: Simon and Schuster, 1984.

———. *The Unconscious God.* New York: Simon and Schuster, 1975.

Geisman, M. Wilma. "The Use in the Missal and the Breviary of the Roman Liturgy of Judith and Esther of the Old Testament in relation to Christian Womanhood." Master's thesis, Notre Dame University, 1956.

Goodier, Alban. "St. Teresa and the Dominicans." *Month.* September 1936. 247–257.

Gospel Parallels: A Synopsis of the First Three Gospels. Edited by Burton H. Throckmorton. 4th ed. Nashville, TN: Thomas Nelson Publishers, 1979.

Gianllanza, Joel. "Spiritual Direction According to St. Teresa of Avila," *Contemplative Review* 12 (Summer 1979), 1–9.

"Hadewijch." By Norbert de Paepe. *New Catholic Encyclopedia.* New York: McGraw Hill, 1967.

Hall, Douglas. *Imaging God.* Grand Rapids, MI: Wm. B. Eerdmans Publishing Co., 1986.

Harrison, Verna. "Male and Female in Cappadocian Theology." *Journal of Theological Studies* n.s. 41:2 (October 1990), 441–471.

Haskins, Susan. *Mary Magdalen: Myth and Metaphor.* New York: Harcourt Brace and Company, 1993.

Hausherr, Irénée. *Spiritual Direction in the Early Christian East.* Translated by A. P. Gythiel. Kalamazoo, MI: Cisterican Publications, 1990.

Hill, Christopher. *World Turned Upside Down.* New York: Penguin Books, 1972.

Holmes, R. M. "Alcoholics Anonymous as Group Logotherapy." *Pastoral Psychology* 21 (1970), 30–36.

Holmes, Urban T. *A History of Christian Spirituality.* New York: Seabury Press, 1980.

Hudson, Winthrop S. "A Suppressed Chapter in Quaker History." *Journal of Religion* 24 (April 1944), 108–118.

Illanes, José Luis. *On the Theology of Work.* Translated by Michael Adams. Dublin: Four Courts Press, 1982.

Johnson, Dale A. *Women in English Religion 1700–1925.* Lewiston, NY: Edwin Mellen Press, 1983.

Kieckhefer, Richard. *Unquiet Souls.* Chicago: University of Chicago Press, 1984.

Kunze, Bonnelyn Young. *Margaret Fell and the Rise of Quakerism.* Stanford, CA: Stanford University Press, 1994.

Ladner, Gerhart B. *The Idea of Reform.* Cambridge, MA: Harvard University Press, 1959.

Lake, Peter. "Feminine Piety and Personal Potency: the 'Emancipation' of Mrs. Jane Ratcliffe." *The Seventeenth Century* 2:2 (July 1987), 143–165.

Laurence, Anne. "A Priesthood of She-Believers: Women and Congregations in Mid-Seventeenth-Century England." In *Women in the Church.* Edited by W. J. Shiels and Diana Wood. 345–363. Oxford: Basil Blackwood, 1990.

Lavrin, Asunciòn. "Unlike Sor Juana? The Model Nun in the Religious Literature of Colonial Mexico." *University of Dayton Review* 16:2 (Spring 1983), 75–92.

Lawrence, C. H. *Medieval Monasticism.* London: Longmans, 1984.

Leech, Kenneth. *Soul Friend: A Study of Spirituality.* London: Sheldon Press, 1977.

Leslie, Robert. *Jesus and Logotherapy.* New York: Abingdon Press, 1965.

Lieban-Kalma, V. "Logotherapy: A Way to Help the Learning Disabled Help Themselves." *Academic Therapy* 19:3 (1984), 262–268.

Life of Henriette d'Osseville (In Religion Mother Ste. Marie) Foundress of the Institute of the Faithful Virgin. Edited by John George Macleod. London: Burns and Oates, 1878.

Lubac, Henri de. *Catholicism.* New York: New American Library, 1964.

Mack, Phyllis. *Visionary Women: Ecstatic Prophecy in Seventeenth-Century England.* Berkeley: University of California Press, 1992.

Mahoney, M. Denis. *Marie of the Incarnation.* Garden City, NY: Doubleday and Company, 1964.

Malingrey, A. M. *Philosophia: Etude d'un groupe de mots dan la litterature grecque.* Paris, 1961.

Martinelli, Serena Spanò. "La biblioteca del 'Corpus Domini' bolognese: l'inconsueto spaccato di una cultura monastica femminile." *La Bibliofilia,* 88:1 (1986), 1–23.

Martos, Joseph. *Doors to the Sacred.* New York: Doubleday and Company, 1981.

McCarthy, Thomas P., comp. *Guide to the Catholic Sisterhoods in the United States.* Washington, DC: Catholic University of America Press, 1955.

McNamara, JoAnn. "The Ordeal of Community: Hagiography and Discipline in Merovingian Convents." *Vox Benedictine* 3 (October 1986), 293–326.

Meister Eckhart and the Beguine Mystics. Edited by Bernard McGinn. New York: Continium, 1994.

Mierlo, Jozef van. *Hadewijch Mengeldichten.* Brussels: n.p., 1912.

Mitchell, G. "Origins of Irish Penance." *Irish Theological Quarterly* 22 (July 1955), 1–14.

Monica, M. *Angela Merici and Her Teaching Idea (1474–1540).* New York: Longman Green and Company, 1927.

Morris, Colin. *The Discovery of the Individual.* London: Camelot Press, 1972.

Murk-Jansen, Saskia. "Hadewijch and Eckhart: Amore intellegere est." In *Meister Eckhart and the Beguine Mystics*. Edited by Bernard McGinn. New York: Continium, 1994.

Murphy, Mary E. "Blessed Clara Gambacorta," Ph.D. dissertation, University of Fribourg, 1928.

Norton, Mary Beth. "My Resting Reaping Times: Sarah Osborn's Defense of Her 'Unfeminine Activities, 1767.'" *Journal of Women in Culture and Society* 2:2 (1976), 515–529.

Novak, Michael. "Controversial Engagements." *First Things* 92 (April 1999), 21–29.

Peifer, Claude. *Monastic Spirituality*. New York: Sheed and Ward, 1966.

Raasch, Juana. "The Monastic Concept of Purity of Heart and its Sources, IV: Early Monasticism." *Studia Monastica* 11 (1969), 269–314.

Rader, Rosemary. "The Role of Celibacy in the Origin and Development of Christian Heterosexual Friendship." Ph.D. dissertation, Stanford University, 1977.

Ranft, Patricia. "The Concept of Witness in the Christian Tradition from Its Origin to Its Institutionalization," *Revue bénédictine* 102: 1–2 (1992), 9–23.

———. "A Key to Counter Reformation Activism: The Confessor-Spiritual Director." *Journal of Feminist Studies in Religion* 10:2 (Fall 1994), 7–26.

———. *Women and Religious Life in Premodern Europe*. New York: St. Martin's Press, 1996.

———. *Women and Spiritual Equality in Christian Tradition*. New York: St. Martin's Press, 1998.

Rapley, Elizabeth. *The Dévotes*. Montreal: McGill-Queen's University Press, 1990.

Raquin, Yves. "The Spiritual Father: Towards Integrating Western and Eastern Spirituality." In *Abba*. Edited by John R. Sommerfeldt. 276–292. Kalamazoo, MI: Cisterican Publishers, 1982.

Rathke, N. N. "Die lutherische Auffassung von der Privatbeichte und ihre Bedeutung für das Kirchliche Leben der Gegenwart." *Monatschrift für Pastoraltheologie* (1917), 29–34.

Reay, Barry. *The Quakers and the English Revolution*. New York: St. Martin's Press, 1985.

Reites, James W. "Ignatius and Ministry with Women." *The Way* supplement 17 (Summer, 1992), 7–19.

Rivet, M. M. *The Influence of the Spanish Mystics on the Works of Saint Francis de Sales*. Washington, DC: Catholic University of America Press, 1941.

Root, Jerry. *"Space to speke": The Confessional Subject in Medieval Literature*. New York: Peter Lang, 1997.

Sargent, George A. "Motivation and Meaning: Frankl's Logotherapy in the Work Situation." Ph.D. dissertation, United States International University, San Diego, 1973.

Semno, George. "Logotherapy in Medical Practice." *International Forum for Logotherapy* 2:2 (Summer/Fall 1979), 12–14.

Shaeffer, Francis A. *True Spirituality*. Wheaton, IL: Tyndale, 1971.

Sheldrake, Philip. *Spirituality and History: Questions of Interpretation and Method.* New York: Crossroad, 1992.

Slade, Carole. *St. Teresa of Avila: Author of a Heroic Life.* Berkeley: University of California Press, 1995.

Stanley, David M., and Raymond E. Brown. "Aspects of New Testament Thought." *Jerome Biblical Commentary.* Edited by Raymond E. Brown, Joseph A. Fitzmeyer, and Roland E. Murphy. Englewood Cliffs, NJ: Prentice-Hall, 1968.

Starck, Patricia. "Rehabilitative Nursing and Logotherapy: A Study of Spinal Cord Injured Clients." *International Forum for Logotherapy* 4:2 (Fall/Winter, 1981), 101–109.

Steinmann, Jean. *Saint Jerome and His Times.* Translated by Ronald Matthew. Notre Dame, IN: Fides Publishers, 1959.

Stopp, Elizabeth. *Madame de Chantal.* Westminster, MD: Newman Press, 1963.

Strong, James. *Strong's Exhaustive Concordance of the Bible.* Nashville, TN: Thomas Nelson Publishers, 1979.

Taylor, Judith C. "From Proselytizing to Social Reform: Three Generations of French Female Teaching Congregations, 1600–1720." Ph.D. dissertation, Arizona State University, 1980.

Trevett, Christine. *Women and Quakerism in the Seventeenth Century.* York: Ebor Press, 1991.

Trites, Allison A. *The Concept of Witness in New Testament Thought.* Cambridge: Cambridge University Press, 1977.

———. "Martus and Martyrdom in the Apocalypse: A Semantic Study." *Novum Testamentum* 15:1 (January 1973), 72–80.

Udovic, Edward. "'Caritas Christi Urgent Nos': The Urgent Challenges of Charity in Seventeenth Century France." *Vincentian Heritage* 12:2 (1991), 85–104.

Vann, Richard T. *The Social Development of English Quakerism, 1655–1755.* Cambridge, MA: Harvard University Press, 1969.

Vawter, Bruce. "The Gospel According to John." In *Jerome Biblical Commentary.* Edited by Raymond E. Brown, Joseph A. Fitzmeyer, and Roland E. Murphy. Englewood Cliffs, NJ: Prentice Hall, 1968.

Vogt, Kari. "'Männ-lichwerden'-Aspekte einer urchristlichen Anthropologie." *Concilium* 21 (1985), 434–442.

Ward, Margaret. *Life of Saint Madeleine Sophie Foundress of the Society of the Sacred Heart of Jesus.* 2nd ed. Roehampton, GB: Convent of the Sacred Heart, 1925.

Ware, Kallistos. "The Spiritual Father in St. John Climacus and St. Symeon the New Theologian." *Studia Patristica* 18 (1988).

Warren, Ann K. *Anchorites and Their Patrons in Medieval England.* Berkeley: University of California Press, 1985.

Weber, Allison. *Teresa of Avila and the Rhetoric of Femininity.* Princeton, NJ: Princeton University Press, 1990.

Weisskopf-Joelson, E. "Paranoia and the Will-to-Meaning." *Existential Psychiatry* 1 (1966), 316–320.

Whidden, M. "Logotherapy in Prison." *International Forum for Logotherapy* 6:11 (Spring/Summer, 1983), 34–39.

Wilcox, Catherine M. *Theology and Women's Ministry in Seventeenth-Century English Quakerism.* Lewiston, NY: Edwin Mellen Press, 1989.

Willard, Dallas. *In Search of Guidance.* Ventura, CA: Regal, 1984.

———. *The Spirit of Disciplines.* San Francisco: Harper San Francisco, 1991.

Williams, Charles E. *The French Oratorians and Absolutism, 1611–1641.* New York: Peter Lang, 1989.

Witnesses for Change. Edited by Elizabeth Potts Brown and Susan Mosher Stuard. New Brunswick, NJ: Rutgers University Press, 1989.

Wright, Wendy. "Jane de Chantal's Guidance of Women." In *Modern Christian Spirituality: Methodology and Historical Essays.* Edited by Bradley C. Hanson. 113–138. Atlanta, GA: Scholars Press, 1990.

Women and Power in the Middle Ages. Edited by Mary Erler and Maryanne Kowlaski. Athens, GA: University of Georgia Press, 1988.

Wood, Jeryldene M. *Women, Art, and Spirituality: The Poor Clares of Early Modern Italy.* Cambridge: Cambridge University Press, 1996.

Index

Printed in the United States
By Bookmasters